Dynamical and Complex Systems

LTCC Advanced Mathematics Series

Series Editors: Shaun Bullett *(Queen Mary University of London, UK)*
Tom Fearn *(University College London, UK)*
Frank Smith *(University College London, UK)*

LTCC Advanced Mathematics Series — Volume 5

Dynamical and Complex Systems

Editors

Shaun Bullett
Queen Mary University of London, UK

Tom Fearn
University College London, UK

Frank Smith
University College London, UK

World Scientific

NEW JERSEY · LONDON · SINGAPORE · BEIJING · SHANGHAI · HONG KONG · TAIPEI · CHENNAI · TOKYO

Published by

World Scientific Publishing Europe Ltd.

57 Shelton Street, Covent Garden, London WC2H 9HE

Head office: 5 Toh Tuck Link, Singapore 596224

USA office: 27 Warren Street, Suite 401-402, Hackensack, NJ 07601

Library of Congress Cataloging-in-Publication Data
Names: Bullett, Shaun, 1947– editor. | Fearn, T., 1949– editor. |
    Smith, F. T. (Frank T.), 1948– editor.
Title: Dynamical and complex systems / [edited by] Shaun Bullett (Queen Mary
    University of London, UK), Tom Fearn (University College London, UK),
    Frank Smith (University College London, UK).
Description: New Jersey : World Scientific, 2017. | Series: LTCC advanced
    mathematics series ; volume 5
Identifiers: LCCN 2016034799 | ISBN 9781786341020 (hc : alk. paper)
Subjects: LCSH: System analysis.
Classification: LCC QA402 .D935 2017 | DDC 515/.39--dc23
LC record available at https://lccn.loc.gov/2016034799

British Library Cataloguing-in-Publication Data
A catalogue record for this book is available from the British Library.

Desk Editors: V. Vishnu Mohan/Mary Simpson

Typeset by Stallion Press
Email: enquiries@stallionpress.com

Preface

The London Taught Course Centre (LTCC) for PhD students in the Mathematical Sciences has the objective of introducing research students to a broad range of topics. For some students, some of these topics might be of obvious relevance to their PhD projects, but the relevance of most will be much less obvious or apparently non-existent. However, all of us involved in mathematical research have experienced that extraordinary moment when the penny drops and some tiny gem of information from outside one's immediate research field turns out to be the key to unravelling a seemingly insoluble problem, or to opening up a new vista of mathematical structure. By offering our students advanced introductions to a range of different areas of mathematics, we hope to open their eyes to new possibilities that they might not otherwise encounter.

Each volume in this series consists of chapters on a group of related themes, based on modules taught at the LTCC by their authors. These modules were already short (five two-hour lectures) and in most cases the lecture notes here are even shorter, covering perhaps three-quarters of the content of the original LTCC course. This brevity was quite deliberate on the part of the editors — we asked the authors to confine themselves to around 35 pages in each chapter, in order to allow as many topics as possible to be included in each volume, while keeping the volumes digestible. The chapters are "advanced introductions", and readers who wish to learn more are encouraged to continue elsewhere. There has been no attempt to make the coverage of topics comprehensive. That would be impossible in any case — any book or series of books which included all that a PhD student in mathematics might need to know would be so large as to be totally unreadable. Instead what we present in this series is a cross-section of some of the topics, both classical and new, that have appeared in LTCC modules in the nine years since it was founded.

The present volume is within the area of dynamical and complex systems. The main readers are likely to be graduate students and more experienced researchers in the mathematical sciences, looking for introductions to areas with which they are unfamiliar. The mathematics presented is intended to be accessible to first year PhD students, whatever their specialised areas of research. Whatever your mathematical background, we encourage you to dive in, and we hope that you will enjoy the experience of widening your mathematical knowledge by reading these concise introductory accounts written by experts at the forefront of current research.

Shaun Bullett, Tom Fearn, Frank Smith

Contents

Chapter 1

Chaos in Statistical Physics

Rainer Klages

School of Mathematical Sciences
Queen Mary University of London
Mile End Road, London E1 4NS, UK

Max Planck Institute for the Physics of Complex Systems
Nöthnitzer Str. 38, D-01187 Dresden, Germany
r.klages@qmul.ac.uk

This chapter introduces to *chaos in dynamical systems* and how this theory can be applied to derive fundamental laws of *statistical physics* from first principles. We first elaborate on the concept of deterministic chaos by defining and calculating Lyapunov exponents and dynamical entropies as fundamental quantities characterising chaos. These quantities are shown to be related to each other by Pesin's theorem. Considering open systems where particles can escape from a set asks for a generalisation of this theorem which involves fractals, whose properties we briefly describe. We then cross-link this theory to statistical physics by discussing simple random walks on the line, their characterisation in terms of diffusion, and the relation to elementary concepts of Brownian motion. This sets the scene for considering the problem of chaotic diffusion. Here we derive a formula exactly expressing diffusion in terms of the chaos quantities mentioned above.

1. Introduction

A dynamical system is a system, represented by points in abstract space, that evolves in time. Very intuitively, one may say that the path of a point particle generated by a dynamical system looks "chaotic" if it displays "random-looking" evolution in time and space. The simplest dynamical systems that can exhibit chaotic dynamics

1

are one-dimensional maps, as we will discuss below. Further details of such dynamics, including a mathematically rigorous definition of chaos, are given in Chapter 6.[a] Surprisingly, abstract low-dimensional chaotic dynamics bears similarities to the dynamics of interacting physical many-particle systems. This is the key point that we explore in this chapter.

Over the past few decades it was found that famous statistical physical laws like Ohm's law for electric conduction, Fourier's law for heat conduction, and Fick's law for the diffusive spreading of particles, which a long time ago were formulated phenomenologically, can be derived *from first principles* in chaotic dynamical systems. This sheds new light on the rigorous mathematical foundations of *Non-equilibrium Statistical Physics*, which is the theory of the dynamics of many-particle systems under external gradients or fields. The external forces induce *transport* of physical quantities like charge, energy, or matter. The goal of non-equilibrium statistical physics is to derive macroscopic statistical laws describing such transport starting from the microscopic dynamics for the single parts of many-particle systems. While the conventional theory puts in randomness "by hand" by using probabilistic, or stochastic, equations of motion like random walks or stochastic differential equations, recent developments in the theory of dynamical systems enable to do such derivations starting from deterministic equations of motion. Determinism means that no random variables are involved, rather, randomness is generated by chaos in the underlying dynamics. This is the field of research that will be introduced by this chapter.

This theory also illustrates the emergence of *complexity* in systems under non-equilibrium conditions: Due to the microscopic nonlinear interaction of the single parts in a complex many-particle system novel, non-trivial dynamics, in this case exemplified by universal transport laws, may emerge on macroscopic scales. The dynamics of a complex system, as a whole, is thus different from the sum of its single parts. In the very simplest case, this idea is illustrated

---

[a]D.K. Arrowsmith, Applied dynamical systems, In *Dynamical and Complex Systems*, eds. S. Bullett, T. Fearn and F. Smith, LTCC Advanced Mathematics Series, Vol. 5, World Scientific, Singapore (2016).

by the interaction of a point particle with a scatterer, where the latter is modelled by a one-dimensional map. This is our vehicle of demonstration in the following, because this simple model can be solved rigorously analytically.

Our chapter consists of two sections: In Section 2 we introduce to two important quantities assessing chaos in dynamical systems, Lyapunov exponents and dynamical entropies. The former are widely used in the applied sciences to test whether a given system is chaotic, the latter is motivated by information theory. Cross-links to *ergodic theory* by defining these quantities are explored, which is a core discipline in mathematical dynamical systems theory. Interestingly, both these different quantities are exactly related to each other by *Pesin's theorem.* Considering *open system* where particles can escape from a set generates *fractals*, a concept that we will introduce as well.

The latter problem cross-links to Section 3, which explores diffusion in chaotic dynamical systems. After briefly introducing to the statistical physical problem of diffusion, we outline a rigorous theory that enables one to calculate diffusion coefficients characterising the spreading of particles from first principles. By combining this approach with a key result for open systems from the previous section, we arrive at an exact formula expressing the diffusion coefficient in terms of the two quantities characterising deterministic chaos introduced before. This important result forms the highlight of our exposition and concludes our chapter.

While Section 2 mainly elaborates on textbook material of chaotic dynamical systems,[1-3] Section 3 introduces to advanced topics that emerged in research over the past 20 years.[4-6]

2. Deterministic Chaos

2.1. *Dynamics of simple maps*

Let us recall the following defintion.

Definition 2.1. Let $J \subseteq \mathbb{R}, x_n \in J, n \in \mathbb{Z}$. Then

$$F : J \to J, \quad x_{n+1} = F(x_n) \tag{2.1}$$

is called a *one-dimensional time-discrete map.* Here $x_{n+1} = F(x_n)$ are sometimes called the *equations of motion* of the dynamical system.

Choosing the initial condition x_0 *determines* the outcome after n discrete time steps, hence we speak of a *deterministic dynamical system*. It works as follows:

$$x_1 = F(x_0) = F^1(x_0)$$

$$x_2 = F(x_1) = F(F(x_0)) = F^2(x_0) \qquad (2.2)$$

$$\Rightarrow \ F^m(x_0) := \underbrace{F \circ F \circ \cdots \circ F(x_0)}_{m\text{-fold composed map}}.$$

In other words, there exists a unique solution to the equations of motion in the form of $x_n = F(x_{n-1}) = \cdots = F^n(x_0)$, which is the counterpart of the flow for time-continuous systems. We will focus on simple piecewise linear maps. The following one serves as a paradigmatic example.[1,2,4,7]

Example 2.2. The Bernoulli shift (also called shift map, doubling map, dyadic transformation).

The map shown in Fig. 1 is defined by

$$B : [0, 1) \to [0, 1), \quad B(x) := 2x \bmod 1 = \begin{cases} 2x, & 0 \le x < 1/2, \\ 2x - 1, & 1/2 \le x < 1. \end{cases}$$
$$(2.3)$$

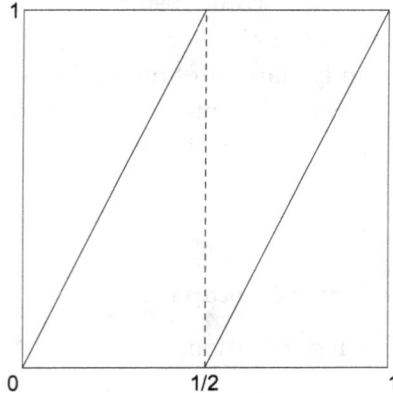

Fig. 1.   The Bernoulli shift.

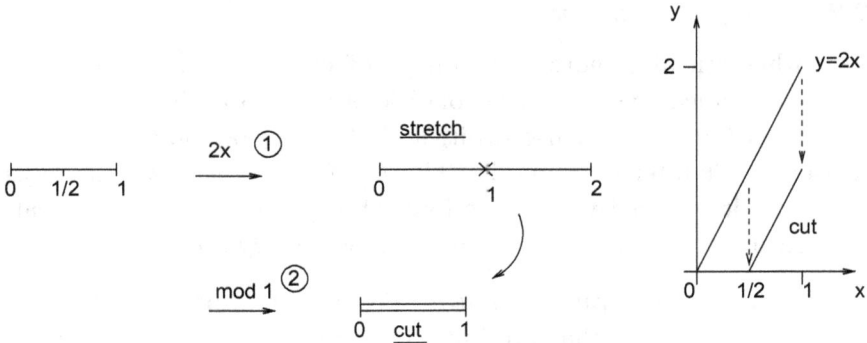

Fig. 2.   Stretch-and-cut mechanism in the Bernoulli shift.

The dynamics of this map can be understood as follows, see Fig. 2: Assume that we fill the whole unit interval with a uniform distribution of points. We may now decompose the action of the Bernoulli shift into two steps:

(1) The map *stretches* the whole distribution of points by a factor of two, which leads to *divergence* of nearby trajectories.
(2) Then we *cut* the resulting line segment in the middle due to the modulo operation mod 1, which leads to motion *bounded* on the unit interval.

The Bernoulli shift thus yields a simple example for an essentially nonlinear stretch-and-cut mechanism, as it typically generates *deterministic chaos*.[2] The same mechanisms are encountered in more realistic dynamical systems. We remark that "stretch and fold" or "stretch, twist and fold" provide alternative mechanisms for generating chaotic behaviour, see, e.g. the tent map mentioned in Chapter 6. The reader may wish to play around with these ideas in thought experiments, where the sets of points is replaced by kneading dough. These ideas can be made mathematically precise by what is called *mixing*, which is an important concept in the ergodic theory of dynamical systems.[4,8]

2.2.   Lyapunov chaos

A mathematically rigorous definition of chaos requires that for a given dynamical system three conditions have to be fulfilled: sensitivity, existence of a dense orbit, and that the periodic points are dense; see Chapter 6 for details. The *Lyapunov exponent* generalises the concept of sensitivity in the form of a quantity that can be calculated more conveniently, as we will motivate by an example.

Example 2.3 (Lyapunov instability of the Bernoulli shift[2]).
Consider two points that are initially displaced from each other by $\delta x_0 := |x_0' - x_0|$ with δx_0 "infinitesimally small" such that x_0, x_0' do not hit different branches of the Bernoulli shift $B(x)$ around $x = 1/2$. We then have

$$\delta x_n := |x_n' - x_n|$$
$$= 2\delta x_{n-1} = 2^2 \delta x_{n-2} = \cdots = 2^n \delta x_0 = e^{n \ln 2} \delta x_0. \quad (2.4)$$

We see that there is an *exponential separation* between two nearby points as we follow their trajectories. The rate of separation $\lambda(x_0) := \ln 2$ is called the (local) *Lyapunov exponent* of the map $B(x)$.

This simple example can be generalised as follows, leading to the general definition of the Lyapunov exponent for one-dimensional maps F. Consider

$$\delta x_n = |x_n' - x_n| = |F^n(x_0') - F^n(x_0)| =: \delta x_0 e^{n\lambda(x_0)} \ (\delta x_0 \to 0) \quad (2.5)$$

for which we *presuppose* that an exponential separation of trajectories exists. By assuming that F is differentiable, we rewrite this equation to

$$\lambda(x_0) = \lim_{n \to \infty} \lim_{\delta x_0 \to 0} \frac{1}{n} \ln \frac{\delta x_n}{\delta x_0}$$
$$= \lim_{n \to \infty} \lim_{\delta x_0 \to 0} \frac{1}{n} \ln \frac{|F^n(x_0 + \delta x_0) - F^n(x_0)|}{\delta x_0}$$
$$= \lim_{n \to \infty} \frac{1}{n} \ln \left| \frac{dF^n(x)}{dx} \right|_{x=x_0}. \quad (2.6)$$

Using the chain rule we obtain

$$\left.\frac{dF^n(x)}{dx}\right|_{x=x_0} = F'(x_{n-1})F'(x_{n-2})\dots F'(x_0), \tag{2.7}$$

which leads to

$$\lambda(x_0) = \lim_{n\to\infty} \frac{1}{n}\ln\left|\prod_{i=0}^{n-1} F'(x_i)\right|$$

$$= \lim_{n\to\infty} \frac{1}{n}\sum_{i=0}^{n-1}\ln\left|F'(x_i)\right|. \tag{2.8}$$

This simple calculation motivates the following definition.

Definition 2.4 (Ref. 7). Let $F \in C^1$ be a map of the real line. The *local Lyapunov exponent* $\lambda(x_0)$ is defined as

$$\lambda(x_0) := \lim_{n\to\infty} \frac{1}{n}\sum_{i=0}^{n-1}\ln\left|F'(x_i)\right|, \tag{2.9}$$

if this limit exists.

Example 2.5. For the Bernoulli shift $B(x) = 2x \bmod 1$ we have $B'(x) = 2$, $\forall x \in [0,1)$, $x \neq \frac{1}{2}$, hence trivially

$$\lambda(x) = \frac{1}{n}\sum_{k=0}^{n-1}\ln 2 = \ln 2 \tag{2.10}$$

at these points.

Note that Definition 2.4 defines the *local* Lyapunov exponent $\lambda(x_0)$, that is, this quantity may depend on our choice of initial conditions x_0. For the Bernoulli shift this is not the case, because this map has a uniform slope of two except at the point of discontinuity, which makes the calculation trivial. Generally, the situation is more complicated. One question is of how to calculate the local Lyapunov exponent, a second one to which extent it depends on initial conditions. An answer to both these questions is provided by the *global* Lyapunov exponent that we are going to introduce, which does not depend on initial conditions and thus characterises the stability of the map as a whole.

It is introduced by observing that the local Lyapunov exponent in Definition 2.4 is defined by a *time average*, where n terms along the trajectory with initial condition x_0 are summed up by averaging over n. That this is not the only possibility to define an average quantity is clarified by the following definition. It requires the concepts of measure and density (see Chapter 6); if the reader is not familiar with these objects, we recommend Ref. 9 as an introduction.

Definition 2.6 (Time and ensemble average[4,8]). Let μ^* be the invariant probability measure of a one-dimensional map F acting on $J \subseteq \mathbb{R}$. Let us consider a function $g : J \to \mathbb{R}$, which we may call an "observable". Then

$$\overline{g(x)} := \lim_{n \to \infty} \frac{1}{n} \sum_{k=0}^{n-1} g(x_k), \qquad (2.11)$$

$x = x_0$, is called the *time* (*or Birkhoff*) *average* of g with respect to F. Now

$$\langle g \rangle := \int_J d\mu^* g(x), \qquad (2.12)$$

where, if such a measure exists, $d\mu^* = \rho^*(x)\,dx$ is called the *ensemble* (*or space*) *average* of g with respect to F. Here $\rho^*(x)$ is the *invariant density* of the map, and $d\mu^*$ is the associated *invariant measure*.[2] Note that $\overline{g(x)}$ may depend on x, whereas $\langle g \rangle$ does not.

If we choose $g(x) = \ln|F'(x)|$ as the observable in Eq. (2.11), we recover Definition 2.4 for the local Lyapunov exponent,

$$\lambda(x) := \overline{\ln|F'(x)|} = \lim_{n \to \infty} \frac{1}{n} \sum_{k=0}^{n-1} \ln|F'(x_k)|, \qquad (2.13)$$

which we may write as $\lambda_t(x) = \lambda(x)$ in order to label it as a time average. If we choose the same observable for the ensemble average Eq. (2.12), we obtain

$$\lambda_e := \langle \ln|F'(x)| \rangle := \int_J dx\, \rho^*(x) \ln|F'(x)|. \qquad (2.14)$$

Example 2.7. For the Bernoulli shift, we have seen that for almost every $x \in [0, 1)$ $\lambda_t = \ln 2$. For λ_e we obtain

$$\lambda_e = \int_0^1 dx \rho^*(x) \ln 2 = \ln 2, \qquad (2.15)$$

taking into account that $\rho^*(x) = 1$ (see Chapter 6). In other words, time and ensemble average are the same for almost every x,

$$\lambda_t(x) = \lambda_e = \ln 2. \qquad (2.16)$$

This motivates the following fundamental definition.

Definition 2.8 (Ergodicity[4,8]). A dynamical system is called *ergodic* if for every g on $J \subseteq \mathbb{R}$ satisfying $\int d\mu^* |g(x)| < \infty$

$$\overline{g(x)} = \langle g \rangle \qquad (2.17)$$

for typical x.

For our purpose it suffices to think of a typical x as a point that is randomly drawn from the invariant density $\rho^*(x)$. This definition implies that for ergodic dynamical systems $\overline{g(x)}$ does not depend on x. That the time average is constant is sometimes also taken as a definition of ergodicity.[3,4] To prove that a given system is ergodic is typically a hard task and one of the fundamental problems in the ergodic theory of dynamical systems; see Refs. 4 and 8 for proofs of ergodicity in case of some simple examples. We remark that pure mathematicians define ergodicity in terms of indecomposability.[10]

On this basis, let us get back to Lyapunov exponents. For time average $\lambda_t(x)$ and ensemble average λ_e of the Bernoulli shift, we have found that $\lambda_t(x) = \lambda_e = \ln 2$. Definition 2.8 now states that the first equality must hold whenever a map F is ergodic. This means, in turn, that for an ergodic dynamical system the Lyapunov exponent becomes a *global* quantity characterising a given map F for a typical point x irrespective of what value we choose for the initial condition, $\lambda_t(x) = \lambda_e = \lambda$. This observation very much facilitates the calculation of λ, as is demonstrated by the following example.

Example 2.9. Let us consider the map $A(x)$ displayed in Fig. 3 below.

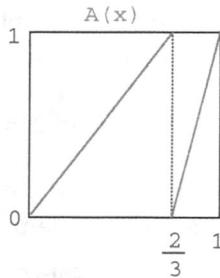

Fig. 3. A simple map for demonstrating the calculation of Lyapunov exponents via ensemble averages.

From the figure we can infer that

$$A(x) := \begin{cases} \dfrac{3}{2}x, & 0 \le x < \dfrac{2}{3}, \\[2mm] 3x - 2, & \dfrac{2}{3} \le x < 1. \end{cases} \qquad (2.18)$$

It is not hard to see that the invariant probability density of this map is uniform, $\rho^*(x) = 1$. The Lyapunov exponent λ for this map is then trivially calculated to

$$\lambda = \int_0^1 dx \rho^*(x) \ln |A'(x)| = \ln 3 - \frac{2}{3} \ln 2. \qquad (2.19)$$

By assuming that map A is ergodic (which here is the case), we can conclude that this result for λ represents the value for typical points in the domain of A.

In other words, for an ergodic map the global Lyapunov exponent λ yields a number that assesses whether it is chaotic in the sense of exhibiting an exponential dynamical instability. This motivates the following definition of deterministic chaos.

Definition 2.10 (Chaos in the sense of Lyapunov[2,3,7]). An ergodic map $F : J \to J$, $J \subseteq \mathbb{R}$, F (piecewise) C^1 is said to be *L-chaotic* on J if $\lambda > 0$.

Why do we introduce a definition of chaos that is different from the rigorous mathematical one discussed in Chapter 6? One reason is that often the largest Lyapunov exponent of a dynamical system is easier to calculate than checking for sensitivity. Furthermore, the magnitude of the positive Lyapunov exponent quantifies the strength of chaos. This is the reason why in the applied sciences "chaos in the sense of Lyapunov" became a very popular concept. Note that there is no unique quantifier of deterministic chaos. Many different definitions are available highlighting different aspects of "chaotic behaviour", all having their advantages and disadvantages. The detailed relations between them are usually non-trivial and a topic of ongoing research. We will encounter yet another definition of chaos in the following section.

2.3. *Entropies*

Let us start with a brief motivation outlining the basic idea of entropy production in dynamical systems. Consider again the Bernoulli shift by decomposing its domain $J = [0, 1)$ into $J_0 := [0, 1/2)$ and $J_1 := [1/2, 1)$. For $x \in [0, 1)$ define the *output map* s (see Chapter 6) by (see Ref. 1)

$$s : [0, 1) \to \{0, 1\}, \quad s(x) := \begin{cases} 0, & x \in J_0, \\ 1, & x \in J_1, \end{cases} \tag{2.20}$$

and let $s_{n+1} := s(x_n)$. Now choose some initial condition $x_0 \in J$. According to the above rule we obtain a digit $s_1 \in \{0, 1\}$. Iterating the Bernoulli shift according to $x_{n+1} = B(x_n)$ then generates a sequence of digits $\{s_1, s_2, \ldots, s_n\}$. This sequence yields nothing else than the binary representation of the given initial condition x_0.[1,2,4] If we assume that we pick an initial condition x_0 at random and feed it into our map without knowing about its precise value, this simple algorithm enables us to find out what number we have actually chosen. In other words, here we have a mechanism of *creation of information* about the initial condition x_0 by analysing the chaotic orbit generated from it as time evolves.

Conversely, if we now assume that we already knew the initial state up to, say, m digits precision and we iterate $p > m$ times, we see that the map simultaneously *destroys information* about the current and future states, in terms of digits, as time evolves. So creation

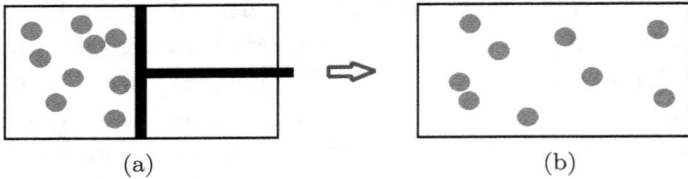

Fig. 4.   Schematic representation of a gas of molecules in a box. In (a) the gas
is constrained by a piston to the left-hand side of the box, in (b) the piston is
removed and the gas can spread out over the whole box. This illustrates the basic
idea of (physical) entropy production.

of information about previous states goes along with loss of information about current and future states. This process is quantified by
the *Kolmogorov–Sinai (KS) entropy* (also called metric, or measure-
theoretic entropy), which measures the exponential rate at which
information is produced, respectively lost in a dynamical system, as
we will see below.

The situation is similar to the following thought experiment illustrated in Fig. 4: Let us assume we have a gas consisting of molecules,
depicted as billiard balls, which is constrained to the left half of the
box as shown in (a). This is like having some information about the
initial conditions of all gas molecules, which are in a more localised,
or ordered, state. If we remove the piston as in (b), we observe that
the gas spreads out over the full box until it reaches a uniform equilibrium steady state. We then have less information available about
the actual positions of all gas molecules, that is, we have increased
the disorder of the whole system. This observation lies at the heart
of what is called *thermodynamic entropy production* in the statistical
physics of many-particle systems which, however, is usually assessed
by quantities that are different from the KS-entropy.

At this point we may not further elaborate on the relation to statistical physical theories. Instead, let us make precise what we mean
by KS-entropy starting from the famous *Shannon (or information)
entropy*.[2,3] This entropy is defined as

$$H_S := \sum_{i=1}^{r} p_i \ln \left(\frac{1}{p_i} \right),$$ (2.21)

where p_i, $i = 1, \ldots, r$, are the probabilities for the r possible outcomes of an experiment. Think, for example, of a roulette game, where carrying out the experiment one time corresponds to $n = 1$ in the iteration of an unknown map. Then H_S measures the amount of uncertainty concerning the outcome of the experiment, which can be understood as follows:

(1) Let $p_1 = 1$, $p_i = 0$ otherwise. By defining $p_i \ln(\frac{1}{p_i}) := 0$, $i \neq 1$, we have $H_S = 0$. This value of the Shannon entropy must therefore characterise the situation where the outcome is completely certain.
(2) Let $p_i = 1/r$, $i = 1, 2, \ldots, r$. Then we obtain $H_S = \ln r$ thus characterising the situation where the outcome is most uncertain because of equal probabilities.

Case (1) thus represents the situation of no information gain by doing the experiment, case (2) corresponds to maximum information gain. These two special cases must therefore define the lower and upper bounds of H_S,

$$0 \leq H_S \leq \ln r. \tag{2.22}$$

This basic concept of information theory carries over to dynamical systems by identifying the probabilities p_i with invariant probability measures μ_i^* on subintervals of a given dynamical system's phase space. The precise connection is worked out in four steps.[2,4]

1. Partition and refinement: Consider a map F acting on $J \subseteq \mathbb{R}$, and let μ^* be an invariant probability measure generated by the map. Let $\{J_i\}$, $i = 1, \ldots, s$, be a *partition* of J.[7] We now construct a *refinement* of this partition as illustrated by the following example.

Example 2.11. Consider the Bernoulli shift displayed in Fig. 5. Start with the partition $\{J_0, J_1\}$ shown in (a). Now create a *refined* partition by iterating these two partition parts backwards according to $B^{-1}(J_i)$ as indicated in (b). Alternatively, you may take the second forward iterate $B^2(x)$ of the Bernoulli shift and then identify the preimages of $x = 1/2$ for this map. In either case the new partition

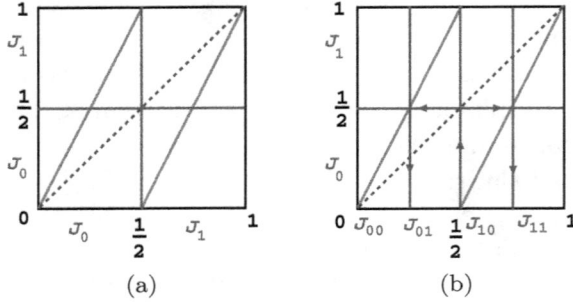

Fig. 5. (a) The Bernoulli shift and a partition of the unit interval consisting of two parts. (b) Refinement of this partition under backward iteration.

parts are obtained to

$$J_{00} := \{x : x \in J_0, \ B(x) \in J_0\},$$
$$J_{01} := \{x : x \in J_0, \ B(x) \in J_1\},$$
$$J_{10} := \{x : x \in J_1, \ B(x) \in J_0\}, \qquad (2.23)$$
$$J_{11} := \{x : x \in J_1, \ B(x) \in J_1\}.$$

If we choose $x_0 \in J_{00}$ we thus know in advance that the orbit emerging from this initial condition under iteration of the map will remain in J_0 at the next iteration. That way, the refined partition clearly yields more information about the dynamics of single orbits.

More generally, for a given map F the above procedure is equivalent to defining

$$\{J_{i_1 i_2}\} := \{J_{i_1} \cap F^{-1}(J_{i_2})\}. \qquad (2.24)$$

The next round of refinement proceeds along the same lines yielding

$$\{J_{i_1 i_2 i_3}\} := \{J_{i_1} \cap F^{-1}(J_{i_2}) \cap F^{-2}(J_{i_3})\}, \qquad (2.25)$$

and so on. For convenience we define

$$\{J_i^n\} := \{J_{i_1 i_2 \dots i_n}\} = \{J_{i_1} \cap F^{-1}(J_{i_2}) \cap \cdots \cap F^{-(n-1)}(J_{i_n})\}. \qquad (2.26)$$

2. H-function: In analogy to the Shannon entropy equation (2.21), next we define the function

$$H(\{J_i^n\}) := - \sum_i \mu^*(J_i^n) \ln \mu^*(J_i^n), \qquad (2.27)$$

where $\mu^*(J_i^n)$ is the invariant measure of the map F on the partition part J_i^n of the nth refinement.

Example 2.12. For the Bernoulli shift with uniform invariant probability density $\rho^*(x) = 1$ and associated (Lebesgue) measure $\mu^*(J_i^n) = \int_{J_i^n} dx\, \rho^*(x) = \text{diam}\,(J_i^n)$ we can calculate

$$H(\{J_i^1\}) = -\left(\frac{1}{2}\ln\frac{1}{2} + \frac{1}{2}\ln\frac{1}{2}\right) = \ln 2$$

$$H(\{J_i^2\}) = H(\{J_{i_1} \cap B^{-1}(J_{i_2})\}) = -4\left(\frac{1}{4}\ln\frac{1}{4}\right) = \ln 4$$

$$H(\{J_i^3\}) = \cdots = \ln 8 = \ln 2^3 \tag{2.28}$$

$$\vdots$$

$$H(\{J_i^n\}) = \ln 2^n.$$

3. Take the limit: We now look at what we obtain in the limit of infinitely refined partition by

$$h(\{J_i^n\}) := \lim_{n\to\infty} \frac{1}{n} H(\{J_i^n\}), \tag{2.29}$$

which defines the rate of gain of information over n refinements.

Example 2.13. For the Bernoulli shift we trivially obtain

$$h(\{J_i^n\}) = \ln 2. \tag{2.30}$$

4. Supremum over partitions: We finish the definition of the KS-entropy by maximising $h(\{J_i^n\})$ over all available partitions,

$$h_{\text{KS}} := \sup_{\{J_i^n\}} h(\{J_i^n\}). \tag{2.31}$$

The last step can be avoided if the partition $\{J_i^n\}$ is *generating* for which it must hold that $\text{diam}\,(J_i^n) \to 0$ $(n \to \infty)$.[3,10] It is quite obvious that for the Bernoulli shift the partition chosen above is generating in that sense, hence $h_{\text{KS}} = \ln 2$ for this map.

These considerations suggest yet another definition of deterministic chaos.

Definition 2.14 (Measure-theoretic chaos[3]). A map $F : J \to J$, $J \subseteq \mathbb{R}$, is said to be chaotic in the sense of exhibiting *dynamical randomness* if $h_{\mathrm{KS}} > 0$.

Again, one may wonder about the relation between this new definition and our previous one in terms of Lyapunov chaos. Let us look again at the Bernoulli shift.

Example 2.15. For $B(x)$ we have calculated the Lyapunov exponent to $\lambda = \ln 2$, see Example 2.7. Above we have seen that $h_{\mathrm{KS}} = \ln 2$ for this map, so we arrive at $\lambda = h_{\mathrm{KS}} = \ln 2$.

That this equality is not an artefact due to the simplicity of our chosen model is stated by the following theorem.

Theorem 2.16 (Pesin's theorem (1977)[4]). *For closed C^2 Anosov systems, the KS-entropy is equal to the sum of positive Lyapunov exponents.*

An Anosov system is a diffeomorphism where the expanding and contracting directions in phase space exhibit a particularly "nice", so-called hyperbolic structure.[4] A proof of this theorem goes considerably beyond the scope of this chapter. In the given formulation, it applies to higher-dimensional dynamical systems that are "suitably well-behaved" in the sense of exhibiting the Anosov property. Applied to one-dimensional maps, it means that if we consider transformations which are "closed" by mapping an interval onto itself, $F : J \to J$, under certain conditions (which we do not further specify here) and if there is a positive Lyapunov exponent $\lambda > 0$ we can expect that $\lambda = h_{\mathrm{KS}}$, as we have seen for the Bernoulli shift. In fact, the Bernoulli shift provides an example of a map that does not fulfil the conditions of the above theorem precisely. However, the theorem can also be formulated under weaker assumptions, and it is believed to hold for an even wider class of dynamical systems.

In order to get an intuition why this theorem should hold, let us look at the information creation in a simple one-dimensional map such as the Bernoulli shift by considering two orbits $\{x_k\}_{k=0}^n$, $\{x_k'\}_{k=0}^n$ starting at nearby initial conditions $|x_0' - x_0| \leq \delta x_0$, $\delta x_0 \ll 1$. Recall the encoding defined by Eq. (2.20). Under the first m iterations these two orbits will then produce the very same sequences of symbols $\{s_k\}_{k=1}^m$, $\{s'_k\}_{k=1}^m$, that is, we cannot distinguish them from each other by our encoding. However, due to the ongoing stretching of the initial displacement δx_0 by a factor of two, eventually there will be an m such that starting from $p > m$ iterations different symbol sequences are generated. Thus we can be sure that in the limit of $n \to \infty$ we will be able to distinguish initially arbitrarily close orbits. If you like analogies, you may think of extracting information about the different initial states via the stretching produced by the iteration process like using a magnifying glass. Therefore, under iteration the exponential rate of separation of nearby trajectories, which is quantified by the positive Lyapunov exponent, must be equal to the rate of information generated, which in turn is given by the KS-entropy. This is at the heart of Pesin's theorem.

2.4. *Open systems, fractals and escape rates*

So far we have only studied closed systems, where intervals are mapped onto themselves. Let us now consider an *open system*, where points can leave the unit interval by never coming back to it. Consequently, in contrast to closed systems the total number of points is not conserved anymore. This situation can be modelled by a slightly generalised example of the Bernoulli shift.

Example 2.17. In the following we will study the map

$$B_a : [0, 1) \to [1 - a/2, a/2), \ B_a(x) := \begin{cases} ax, & 0 \leq x < 1/2, \\ ax + 1 - a, & 1/2 \leq x < 1, \end{cases}$$
$$(2.32)$$

see Fig. 6, where the slope $a \geq 2$ defines a control parameter. For $a = 2$ we recover our familiar Bernoulli shift, whereas for $a > 2$ the map defines an open system. That is, whenever points are mapped into the escape region of width Δ these points are removed from the unit interval. You may thus think of the escape region as a subinterval that absorbs any particles mapped onto it.

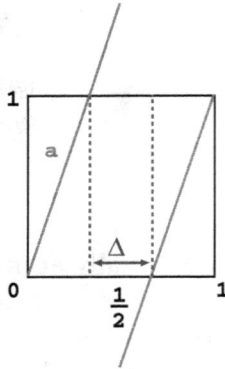

Fig. 6. A generalisation of the Bernoulli shift, defined as a parameter-dependent map $B_a(x)$ modelling an open system. The slope a defines a control parameter, Δ denotes the width of the escape region.

We now wish to compute the number of points N_n remaining on the unit interval at time step n, where we start from a uniform distribution of $N_0 = N$ points on this interval at $n = 0$. This can be done as follows: Recall that the probability density $\rho_n(x)$ was defined by

$$\rho_n(x) := \frac{N_{n,j}}{N dx}, \tag{2.33}$$

where $N_{n,j}$ is the number of points in the interval dx centred around the position x_j at time step n (see Chapter 6). With $N_n = \sum_j N_{n,j}$ we have that

$$N_1 = N_0 - \rho_0 N \Delta. \tag{2.34}$$

By observing that for $B_a(x)$, starting from $\rho_0 = 1$ points are always uniformly distributed on the unit interval at subsequent iterations, we can derive an equation for the density ρ_1 of points covering the unit interval at the next time step $n = 1$. For this purpose, we divide the above equation by the total number of points N (multiplied with the total width of the unit interval, which however is one), which yields

$$\rho_1 = \frac{N_1}{N} = \rho_0 - \rho_0 \Delta = \rho_0 (1 - \Delta). \tag{2.35}$$

This procedure can be reiterated starting now from

$$N_2 = N_1 - \rho_1 N \Delta \qquad (2.36)$$

leading to

$$\rho_2 = \frac{N_2}{N} = \rho_1(1 - \Delta), \qquad (2.37)$$

and so on. For general n we thus obtain

$$\rho_n = \rho_{n-1}(1 - \Delta) = \rho_0(1 - \Delta)^n = \rho_0 e^{n \ln(1-\Delta)}, \qquad (2.38)$$

or correspondingly

$$N_n = N_0 e^{n \ln(1-\Delta)}, \qquad (2.39)$$

which suggests the following definition.

Definition 2.18. For an open system with exponential decrease of the number of points,

$$N_n = N_0 e^{-\gamma n}, \qquad (2.40)$$

γ is called the *escape rate*.

In case of our mapping we thus identify

$$\gamma = \ln \frac{1}{1 - \Delta} \qquad (2.41)$$

as the escape rate. We may now wonder whether there are any initial conditions that never leave the unit interval and about the character of this set of points. The set can be constructed as exemplified for $B_a(x)$, $a = 3$, in Fig. 7.

Example 2.19. Let us start again with a uniform distribution of points on the unit interval. We can then see that the points which remain on the unit interval after one iteration of the map form two sets, each of length $1/3$. Iterating now the boundary points of the escape region backwards in time according to $x_n = B_3^{-1}(x_{n+1})$, we can obtain all preimages of the escape region. We find that initial points which remain on the unit interval after two iterations belong to four smaller sets, each of length $1/9$, as depicted at the bottom of Fig. 7. Repeating this procedure infinitely many times reveals that

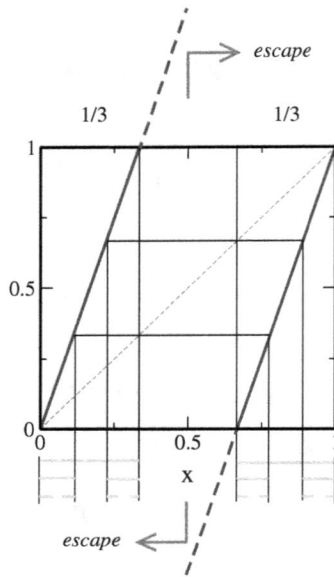

Fig. 7. Construction of the set $\mathcal{C}_{B_3}$ of initial conditions of the map $B_3(x)$ that never leave the unit interval.

the points which never leave the unit interval form the very special set $\mathcal{C}_{B_3}$, which is known as the *middle third Cantor set*.

Definition 2.20 (Cantor set[2]). A *Cantor set* is a closed set which consists entirely of boundary points each of which is a limit point of the set.

Let us explore some fundamental properties of the set $\mathcal{C}_{B_3}$ (see Ref. 2):

(1) From Fig. 7 we can infer that the total length l_n of the intervals of points remaining on the unit interval after n iterations, which is identical with the Lebesgue measure μ_L of these sets, is

$$l_0 = 1, \quad l_1 = \frac{2}{3}, \quad l_2 = \frac{4}{9} = \left(\frac{2}{3}\right)^2, \ldots, l_n = \left(\frac{2}{3}\right)^n. \quad (2.42)$$

We thus see that

$$l_n = \left(\frac{2}{3}\right)^n \to 0 \; (n \to \infty), \quad (2.43)$$

that is, the total length of this set goes to zero, $\mu_L(\mathcal{C}_{B_3}) = 0$. However, there exist also Cantor sets whose Lebesgue measure is larger than zero.[2] Note that matching $l_n = \exp(-n \ln(3/2))$ to Eq. (2.41) yields an escape rate of $\gamma = \ln(3/2)$ for this map.

(2) By using the binary encoding equation (2.20) for all intervals of $\mathcal{C}_{B_3}$, thus mapping all elements of this set onto all the numbers in the unit interval, it can nevertheless be shown that our Cantor set contains an *uncountable number of points*.[4]

(3) By construction $\mathcal{C}_{B_3}$ must be the invariant set of the map $B_3(x)$ under iteration, so the invariant measure of our open system must be the measure defined on the Cantor set, $\mu^*(\mathcal{C})$, $\mathcal{C} \in \mathcal{C}_{B_3}$; see Ref. 10, and also Example 2.23 for the procedure of how to calculate this measure.

(4) For the next property we need the following definition.

Definition 2.21 (Repeller[3,4]). The limit set of points that never escape is called a *repeller*. The orbits that escape are *transients*, and $1/\gamma$ is the typical duration of them.

From this we can conclude that $\mathcal{C}_{B_3}$ represents the repeller of the map $B_3(x)$.

(5) Since $\mathcal{C}_{B_3}$ is completely disconnected by only consisting of boundary points, its topology is highly singular. Consequently, no invariant density $\rho^*(x)$ can be defined on this set, since this concept presupposes a certain "smoothness" of the underlying topology such that one can meaningfully speak of "small subintervals dx" on which one counts the number of points, see Eq. (2.33). In contrast, $\mu^*(\mathcal{C})$ is still well-defined, and we speak of it as a *singular measure*.[4]

(6) Figure 7 shows that $\mathcal{C}_{B_3}$ is *self-similar*, in the sense that smaller pieces of this structure reproduce the entire set upon magnification.[2] Here we find that the whole set can be reproduced by magnifying the fundamental structure of two subsets with a gap in the middle by a constant factor of three. Often such a simple scaling law does not exist for these types of sets. Instead, the scaling may depend on the position x of the subset, in which case one speaks of a *self-affine* structure.[2,3,7]

(7) Again we need a definition.

Definition 2.22 (Fractals, qualitatively[3]). *Fractals* are geometrical objects that possess non-trivial structure on arbitrarily fine scales.

In case of our Cantor set $\mathcal{C}_{B_3}$, these structures are generated by a simple scaling law. However, generally fractals can be arbitrarily complicated on finer and finer scales. An example of a structure that is trivial, hence not fractal, is a straight line. The fractality of such complicated sets can be assessed by quantities called *fractal dimensions*,[2,3] which generalise the integer dimensionality of Euclidean geometry. It is interesting how in our case fractal geometry naturally comes into play, forming an important ingredient of the theory of dynamical systems. However, here we do not further elaborate on the concept of fractal geometry and refer to the literature instead.[2,3,7]

Example 2.23. Let us now compute all three basic quantities that we have introduced so far, that is: the Lyapunov exponent λ and the KS-entropy h_{ks} *on* the invariant set as well as the escape rate γ *from* this set. We do so for the map $B_3(x)$ which, as we have learned, produces a fractal repeller. According to Eqs. (2.12) and (2.14) we have to calculate

$$\lambda(\mathcal{C}_{B_3}) = \int_0^1 d\mu^* \ln |B_3'(x)|. \qquad (2.44)$$

However, for typical points we have $B_3'(x) = 3$, hence the Lyapunov exponent must trivially be

$$\lambda(\mathcal{C}_{B_3}) = \ln 3, \qquad (2.45)$$

because the probability measure μ^* is normalised. The calculation of the KS-entropy requires a bit more work: Recall that

$$H(\{\mathcal{C}_i^n\}) := - \sum_{i=1}^{2^n} \mu^*(\mathcal{C}_i^n) \ln \mu^*(\mathcal{C}_i^n), \qquad (2.46)$$

see Eq. (2.27), where $\mathcal{C}_i^n$ denotes the ith part of the emerging Cantor set at the nth level of its construction. We now proceed along the lines of Example 2.12. From Fig. 7 we can infer that

$$\mu^*(C_i^1) = \frac{\frac{1}{3}}{\frac{2}{3}} = \frac{1}{2}$$

at the first level of refinement. Note that here we have *renormalised* the (Lebesgue) measure on the partition part C_i^1. That is, we have divided the measure by the total measure surviving on all partition parts such that we always arrive at a proper probability measure under iteration. The measure constructed that way is known as the *conditionally invariant measure* on the Cantor set.[3] Repeating this procedure yields

$$\mu^*(C_i^2) = \frac{\frac{1}{9}}{\frac{4}{9}} = \frac{1}{4}$$

$$\vdots \tag{2.47}$$

$$\mu^*(C_i^n) = \frac{\left(\frac{1}{3}\right)^n}{\left(\frac{2}{3}\right)^n} = 2^{-n}$$

from which we obtain

$$H(\{C_i^n\}) = -\sum_{i=1}^{2^n} 2^{-n} \ln 2^{-n} = n \ln 2. \tag{2.48}$$

We thus see that by taking the limit according to Eq. (2.29) and noting that our partitioning is generating on the fractal repeller $\mathcal{C}_{B_3} = \{C_i^\infty\}$, we arrive at

$$h_{\mathrm{KS}}(\mathcal{C}_{B_3}) = \lim_{n\to\infty} \frac{1}{n} H(\{C_i^n\}) = \ln 2. \tag{2.49}$$

Finally, with Eq. (2.41) and an escape region of size $\Delta = 1/3$ for $B_3(x)$ we get for the escape rate

$$\gamma(\mathcal{C}_{B_3}) = \ln \frac{1}{1-\Delta} = \ln \frac{3}{2}, \tag{2.50}$$

as we have already seen before.

In summary, we have that $\gamma(\mathcal{C}_{B_3}) = \ln\frac{3}{2} = \ln 3 - \ln 2$, $\lambda(\mathcal{C}_{B_3}) = \ln 3$, $h_{\mathrm{KS}}(\mathcal{C}_{B_3}) = \ln 2$, which suggests the relation

$$\gamma(\mathcal{C}_{B_3}) = \lambda(\mathcal{C}_{B_3}) - h_{\mathrm{KS}}(\mathcal{C}_{B_3}). \tag{2.51}$$

Again, this equation is no coincidence. It is a generalisation of Pesin's theorem to open systems, known as the *escape rate formula*. This equation holds under similar conditions like Pesin's theorem, which is recovered from it if there is no escape.[4]

3. Chaotic Diffusion

We now apply the concepts of dynamical systems theory developed up to now to a fundamental problem in non-equilibrium statistical physics, which is to understand the microscopic origin of diffusion in many-particle systems. We start with a reminder of diffusion as a simple random walk on the line. Modelling such processes by suitably generalising the piecewise linear map studied previously, we will see how diffusion can be generated by microscopic deterministic chaos. The main result will be an exact formula relating the diffusion coefficient, which characterises macroscopic diffusion of particles, to the dynamical systems quantities introduced before.

3.1. *What is chaotic diffusion?*

In order to learn about chaotic diffusion, we must first understand what ordinary diffusion is all about. Here we introduce this concept by means of a famous example, see Fig. 8: Let us imagine that some evening a sailor wants to walk home, however, he is completely drunk such that he has no control over his single steps. For sake of simplicity let us imagine that he moves in one dimension. He starts at a lamppost at position $x = 0$ and then makes steps of a certain step length s to the left and to the right. Since he is completely drunk he loses all memory between any single steps, that is, all steps are *uncorrelated*. It is like tossing a coin in order to decide whether to go to the left or to the right at the next step. We may now ask for the probability to find the sailor after n steps at position x, i.e. a distance $|x|$ away from his starting point.

The answer to this question is obtained from a calculation for an ensemble of sailors starting from the lamppost and is given in terms of Gaussian probability distributions for the sailor's positions,

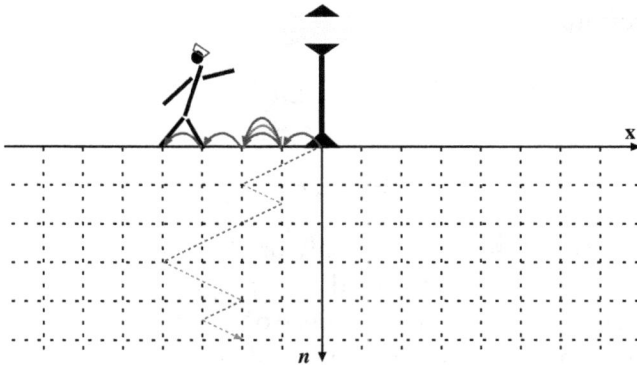

Fig. 8. The "problem of the random walk" in terms of a drunken sailor at a lamppost. The space-time diagram shows an example of a trajectory for such a drunken sailor, where $n \in \mathbb{N}$ holds for discrete time and $x \in \mathbb{R}$ for the position of the sailor on a discrete lattice of spacing s.

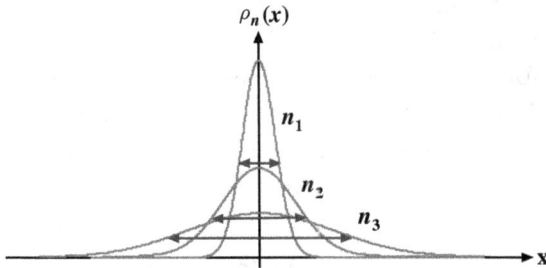

Fig. 9. Probability distribution functions $\rho_n(x)$ to find a sailor after n time steps at position x on the line, calculated for an ensemble of sailors starting at the lamppost, cf. Fig. 8. Shown are three probability densities after different numbers of iteration $n_1 < n_2 < n_3$.

which are obtained in a suitable scaling limit.[11] Figure 9 sketches the spreading of such a diffusing distribution of sailors in time. The mathematical reason for the emerging Gaussianity of the probability distributions is nothing else than the central limit theorem.

We may now wish to quantify the speed by which a "droplet of sailors" starting at the lamppost spreads out. This can be done by calculating the *diffusion coefficient* for this system. In case of one-dimensional dynamics the diffusion coefficient can be defined by the

Einstein formula

$$D := \lim_{n \to \infty} \frac{1}{2n} \langle x^2 \rangle, \tag{3.1}$$

where

$$\langle x^2 \rangle := \int dx \; x^2 \rho_n(x) \tag{3.2}$$

is the variance, or second moment, of the probability distribution $\rho_n(x)$ at time step n, also called *mean square displacement* of the particles. This formula may be understood as follows: For our ensemble of sailors we may choose $\rho_0(x) = \delta(x)$ as the initial probability distribution with $\delta(x)$ denoting the (Dirac) δ-function, which mimics the situation that all sailors start at the same lamppost at $x = 0$. If our system is ergodic, the diffusion coefficient should be independent of the choice of the initial ensemble. The spreading of the distribution of sailors is then quantified by the growth of the mean square displacement in time. If this quantity grows linearly in time, which may not necessarily be the case but holds true if our probability distributions for the positions are Gaussian in the long-time limit,[6] the magnitude of the diffusion coefficient D tells us how quickly our ensemble of sailors disperses. For further details about a statistical physics description of diffusion we refer to Ref. 11.

In contrast to this well-known picture of diffusion as a stochastic random walk, the theory of dynamical systems makes it possible to treat diffusion as a *deterministic dynamical process*. Let us replace the sailor by a point particle. Instead of coin tossing, the orbit of such a particle starting at initial condition x_0 may then be generated by a *chaotic* dynamical system of the type as considered in the previous sections, $x_{n+1} = F(x_n)$. Note that defining the one-dimensional map $F(x)$ together with this equation yields the *full microscopic equations of motion* of the system. You may think of these equations as a caricature of Newton's equations of motion modelling the diffusion of a single particle. Most importantly, in contrast to the drunken sailor with his memory loss after any time step here the *complete memory* of a particle is taken into account, that is, all steps are fully correlated. The decisive new fact that distinguishes this dynamical process from the one of a simple uncorrelated random walk is hence that x_{n+1} is

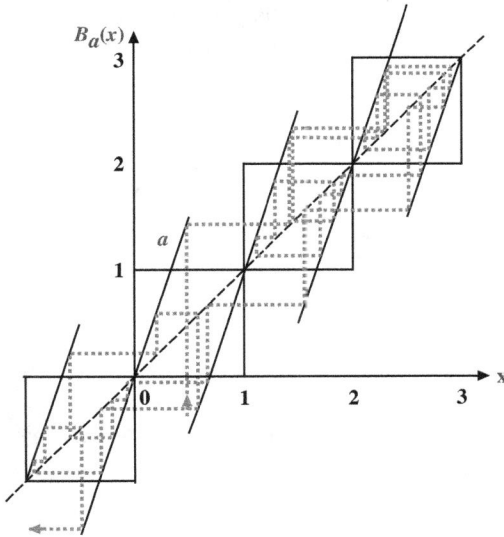

Fig. 10.   A simple model for chaotic diffusion. The dashed line depicts the orbit of a diffusing particle in the form of a cobweb plot.[7] The slope a serves as a control parameter for the periodically continued piecewise linear map $B_a(x)$.

uniquely determined by x_n, rather than having a random distribution of x_{n+1} for a given x_n. If the resulting dynamics of an ensemble of particles for given equations of motion has the property that a diffusion coefficient $D > 0$ Eq. (3.1) exists, we speak of *deterministic* or *chaotic diffusion*.[1,4–6]

Figure 10 shows the simple model of chaotic diffusion that we shall study in the following. It depicts a "chain of boxes" of chain length $L \in \mathbb{N}$, which continues periodically in both directions to infinity, and the orbit of a moving point particle. Let us first specify the map defined on the unit interval, which we may call the box map. For this we choose the map $B_a(x)$ introduced in Example 2.17. We can now periodically continue this box map onto the whole real line by a *lift of degree one*,

$$B_a(x + 1) = B_c(x) + 1. \tag{3.3}$$

Physically speaking, this means that $B_a(x)$ continued onto the real line is translational invariant with respect to integers. Note furthermore that we have chosen a box map whose graph is point symmetric with respect to the centre of the box at $(x, y) = (0.5, 0.5)$. This implies that the graph of the full map $B_a(x)$ is anti-symmetric with respect to $x = 0$,

$$B_a(x) = -B_a(-x), \tag{3.4}$$

so that there is no "drift" in this chain of boxes. The drift case with broken symmetry could be studied as well,[6] but we exclude it here for sake of simplicity.

3.2.   *The diffusion equation*

In the last section we have sketched in a nutshell what, in our setting, we mean if we speak of diffusion. This picture is made more precise by deriving an equation that exactly generates the dynamics of the probability densities displayed in Fig. 9.[11] For this purpose, let us reconsider for a moment the situation depicted in Fig. 4. There, we had a gas with an initially very high concentration of particles on the left-hand side of the box. After the piston was removed, it seemed natural that the particles spread out over the right-hand side of the box as well thus diffusively covering the whole box. We may thus come to the conclusion that, firstly, *there will be diffusion if the density of particles in a substance is non-uniform in space.* For this density of particles and by restricting ourselves to diffusion in one dimension in the following, let us write $\tilde{n} = \tilde{n}(x, t)$, which holds for the number of particles that we can find in a small line element dx around the position x at time step t divided by the total number of particles N.

As a second observation, we see that *diffusion occurs in the direction of decreasing particle density.* This may be expressed as

$$j =: -D\frac{\partial \tilde{n}}{\partial x}, \tag{3.5}$$

which according to Einstein's formula Eq. (3.1) may be considered as a second definition of the diffusion coefficient D. Here the flux $j = j(x, t)$ denotes the number of particles passing through an area perpendicular to the direction of diffusion per time t. This equation is known as *Fick's first law.* Finally, let us assume that no particles are created or destroyed during our diffusion process. In other words, we have *conservation of the number of particles* in the form of

$$\frac{\partial \tilde{n}}{\partial t} + \frac{\partial j}{\partial x} = 0. \tag{3.6}$$

This *continuity equation* expresses the fact that whenever the particle density $\tilde{n}$ changes in time t, it must be due to a spatial change in the particle flux j. Combining the equation with Fick's first law, we obtain *Fick's second law*,

$$\frac{\partial \tilde{n}}{\partial t} = D \frac{\partial^2 \tilde{n}}{\partial x^2}, \tag{3.7}$$

which is also known as the *diffusion equation*. Mathematicians call the process defined by this equation a *Wiener process*, whereas physicists rather speak of *Brownian motion*. If we would now solve the diffusion equation for the drunken sailor initial density $\tilde{n}(x,0) = \delta(x)$, we would obtain the precise functional form of our spreading Gaussians in Fig. 9,

$$\tilde{n}(x,t) = \frac{1}{\sqrt{4\pi Dt}} \exp\left(-\frac{x^2}{4Dt}\right). \tag{3.8}$$

Calculating the second moment of this distribution according to Eq. (3.2) would lead us to recover Einstein's definition of the diffusion coefficient Eq. (3.1). Therefore, both this definition and the one provided by Fick's first law are consistent with each other.

3.3. Basics of the escape rate formalism

We are now fully prepared for establishing an interesting link between dynamical systems theory and statistical mechanics. We start with a brief outline of the concept of this theory, which is called the *escape rate formalism*.[4,5] It consists of three steps.

Step 1: *Solve the one-dimensional diffusion equation* (3.7) *derived above for absorbing boundary conditions.* That is, we consider now some type of *open* system similar to what we have studied in the previous section. We may thus expect that the total number of particles $N(t) := \int dx\, \tilde{n}(x,t)$ within the system decreases exponentially as time evolves according to the law expressed by Eq. (2.40), that is,

$$N(t) = N(0)e^{-\gamma_{\text{de}}t}. \tag{3.9}$$

It will turn out that the escape rate γ_{de} defined by the diffusion equation with absorbing boundaries is a function of the system size L and of the diffusion coefficient D.

Step 2: *Solve the Frobenius–Perron equation*

$$\rho_{n+1}(x) = \int dy\, \rho_n(y)\, \delta(x - F(y)), \tag{3.10}$$

which represents the continuity equation for the probability density $\rho_n(x)$ of the map $F(x)$ (see Chapter 6 and Refs. 2–4), for the very same *absorbing boundary conditions* as in Step 1. Let us assume that the dynamical system under consideration is normal diffusive, that is, that a diffusion coefficient $D > 0$ exists. We may then expect a decrease in the number of particles that is completely analogous to what we have obtained from the diffusion equation. That is, if we define as before $N_n := \int dx\, \rho_n(x)$ as the total number of particles within the system at discrete time step n, in case of normal diffusion we should obtain

$$N_n = N_0 e^{-\gamma_{\text{FP}}n}. \tag{3.11}$$

However, in contrast to Step 1 here the escape rate γ_{FP} should be fully determined by the dynamical system that we are considering. In fact, we have already seen before that for open systems the escape rate can be expressed exactly as the difference between the positive Lyapunov exponent and the KS-entropy on the fractal repeller, cf. the escape rate formula Eq. (2.51).

Step 3: If the functional forms of the particle density $\tilde{n}(x,t)$ of the diffusion equation and of the probability density $\rho_n(x)$ of the map's Frobenius–Perron equation *match in the limit of system size and time going to infinity* — which is what one has to show —, the escape rates γ_{de} obtained from the diffusion equation and γ_{FP} calculated from the Frobenius–Perron equation should be equal,

$$\gamma_{\text{de}} = \gamma_{\text{FP}}, \tag{3.12}$$

providing a fundamental link between the statistical physical theory of diffusion and dynamical systems theory. Since γ_{de} is a function of the diffusion coefficient D, and knowing that γ_{FP} is a function of dynamical systems quantities, we should then be able to express D exactly in terms of these dynamical systems quantifiers. We will now illustrate how this method works by applying it to our simple chaotic diffusive model introduced above.

3.4. *The escape rate formalism applied to a simple map*

Let us consider the map $B_a(x)$ lifted onto the whole real line for the specific parameter value $a = 4$, see Fig. 11. With L we denote the chain length. Proceeding along the above lines, let us start with the following steps.

Step 1: Solve the one-dimensional diffusion equation (3.7) for the *absorbing boundary conditions*

$$\tilde{n}(0, t) = \tilde{n}(L, t) = 0, \tag{3.13}$$

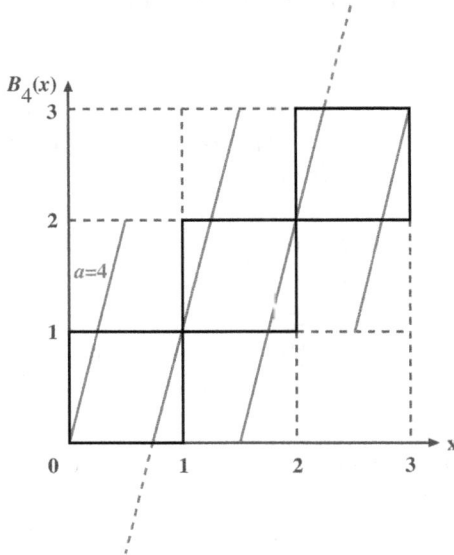

Fig. 11. Our previous map $B_a(x)$ periodically continued onto the whole real line for the specific parameter value $a = 4$. The example shown depicts a chain of length $L = 3$. The dashed quadratic grid indicates a Markov partition for this map.

which models the situation that particles escape precisely at the boundaries of our one-dimensional domain. A straightforward calculation yields

$$\tilde{n}(x,t) = \sum_{m=1}^{\infty} b_m \exp\left(-\left(\frac{m\pi}{L}\right)^2 Dt\right) \sin\left(\frac{m\pi}{L}x\right) \qquad (3.14)$$

with b_m denoting the Fourier coefficients.

Step 2: Solve the Frobenius–Perron equation (3.10) for the same absorbing boundary conditions,

$$\rho_n(0) = \rho_n(L) = 0. \qquad (3.15)$$

In order to do so, we first need to introduce *Markov partitions* for our map $B_4(x)$.

Definition 3.1 (Markov partition, verbally[3]). For one-dimensional maps acting on compact intervals a partition is called *Markov* if *parts of the partition* get mapped again onto *parts of the partition*, or onto *unions of parts of the partition*.

Example 3.2. The dashed quadratic grid in Fig. 11 defines a Markov partition for the lifted map $B_4(x)$.

Having a Markov partition at hand enables us to rewrite the Frobenius–Perron equation in the form of a matrix equation, where a Frobenius–Perron matrix operator acts onto probability density vectors defined with respect to this special partitioning. In order to see this, consider an initial density of points that covers, e.g. the interval in the second box of Fig. 11 uniformly. By applying the map onto this density, one observes that points of this interval get mapped twofold onto the interval in the second box again, but that there is also escape from this box which uniformly covers the third and the first box intervals, respectively. This mechanism applies to any box in our chain of boxes, modified only by the absorbing boundary conditions at the ends of the chain of length L. Taking into account the stretching of the density by the slope $a = 4$ at each iteration, this suggests that the Frobenius–Perron equation (3.10) can be rewritten as

$$\rho_{n+1} = \frac{1}{4} T(4)\,\rho_n, \tag{3.16}$$

where the $L \times L$-transition matrix $T(4)$ must read

$$T(4) = \begin{pmatrix} 2 & 1 & 0 & 0 & \cdots & 0 & 0 & 0 \\ 1 & 2 & 1 & 0 & 0 & \cdots & 0 & 0 \\ 0 & 1 & 2 & 1 & 0 & 0 & \cdots & 0 \\ \vdots & & & \vdots & \vdots & & & \vdots \\ 0 & \cdots & 0 & 0 & 1 & 2 & 1 & 0 \\ 0 & 0 & \cdots & 0 & 0 & 1 & 2 & 1 \\ 0 & 0 & 0 & \cdots & 0 & 0 & 1 & 2 \end{pmatrix}. \tag{3.17}$$

Note that in any row and in any column we have three non-zero matrix elements except in the very first and the very last rows and columns, which reflect the absorbing boundary conditions. In Eq. (3.16) this transition matrix $T(4)$ is applied to a column vector ρ_n corresponding to the probability density $\rho_n(x)$, which can be written as

$$\rho_n = |\rho_n(x)\rangle := (\rho_n^1, \rho_n^2, \ldots, \rho_n^k, \ldots, \rho_n^L)^*, \tag{3.18}$$

where "$*$" denotes the transpose and ρ_n^k represents the component of the probability density in the kth box, $\rho_n(x) = \rho_n^k$, $k - 1 < x \le k$, $k = 1, \ldots, L$, ρ_n^k being constant on each part of the partition. We see that this transition matrix is symmetric, hence it can be diagonalised by spectral decomposition. Solving the eigenvalue problem

$$T(4)\,|\phi_m(x)\rangle = \chi_m(4)\,|\phi_m(x)\rangle, \tag{3.19}$$

where $\chi_m(4)$ and $|\phi_m(x)\rangle$ are the eigenvalues and eigenvectors of $T(4)$, respectively, one obtains

$$|\rho_n(x)\rangle = \frac{1}{4} \sum_{m=1}^{L} \chi_m(4)\,|\phi_m(x)\rangle\langle\varphi_m(x)|\rho_{n-1}(x)\rangle$$

$$= \sum_{m=1}^{L} \exp\left(-n \ln \frac{4}{\chi_m(4)}\right) |\phi_m(x)\rangle\langle\phi_m(x)|\rho_0(x)\rangle, \tag{3.20}$$

where $|\rho_0(x)\rangle$ is the initial probability density vector. Note that the choice of initial probability densities is restricted by this method to functions that can be written in the vector form of Eq. (3.18). It remains to solve the eigenvalue problem (3.19).[6] The eigenvalue equation for the single components of the matrix $T(4)$ reads

$$\phi_m^k + 2\phi_m^{k+1} + \phi_m^{k+2} = \chi_m \phi_m^{k+1}, \quad 0 \le k \le L-1, \tag{3.21}$$

supplemented by the absorbing boundary conditions

$$\phi_m^0 = \phi_m^{L+1} = 0. \tag{3.22}$$

This equation is the form of a discretised ordinary differential equation of degree two, hence we make the ansatz

$$\phi_m^k = a\cos(k\theta) + b\sin(k\theta), \quad 0 \le k \le L+1. \tag{3.23}$$

The two boundary conditions lead to

$$a = 0 \quad \text{and} \quad \sin((L+1)\theta) = 0 \tag{3.24}$$

yielding

$$\theta_m = \frac{m\pi}{L+1}, \quad 1 \le m \le L. \tag{3.25}$$

The eigenvectors are then determined by

$$\phi_m^k = b\sin(k\theta_m). \tag{3.26}$$

Combining this equation with Eq. (3.21) yields as the eigenvalues

$$\chi_m = 2 + 2\cos\theta_m. \tag{3.27}$$

Step 3: Putting all details together, it remains to match the solution of the diffusion equation to the one of the Frobenius–Perron equation: In the limit of time t and system size L to infinity, the density $\tilde{n}(x,t)$ (Eq. (3.14)) of the diffusion equation reduces to the largest eigenmode,

$$\tilde{n}(x,t) \simeq \exp\left(-\gamma_{\text{de}}t\right) B\sin\left(\frac{\pi}{L}x\right), \tag{3.28}$$

where

$$\gamma_{\text{de}} := \left(\frac{\pi}{L}\right)^2 D \tag{3.29}$$

defines the escape rate as determined by the diffusion equation. Analogously, for discrete time n and chain length L to infinity we obtain for the probability density of the Frobenius–Perron equation, Eq. (3.20) with Eq. (3.26),

$$\rho_n(x) \simeq \exp\left(-\gamma_{\text{FP}} n\right) \tilde{B} \sin\left(\frac{\pi}{L+1} k\right),$$

$$k = 0, \ldots, L+1. \quad k-1 < x \leq k \qquad (3.30)$$

with an escape rate of this dynamical system given by

$$\gamma_{\text{FP}} = \ln \frac{4}{2 + 2\cos(\pi/(L+1))}, \qquad (3.31)$$

which is determined by the largest eigenvalue χ_1 of the matrix $T(4)$, see Eq. (3.20) with Eq. (3.27). We can now see that the functional forms of the eigenmodes of Eqs. (3.28) and (3.30) match precisely. This allows us to match Eqs. (3.29) and (3.31) leading to

$$D(4) = \left(\frac{L}{\pi}\right)^2 \gamma_{\text{FP}}. \qquad (3.32)$$

Using the right-hand side of Eq. (3.31) and expanding it for $L \to \infty$, this formula enables us to calculate the diffusion coefficient $D(4)$ to

$$D(4) = \left(\frac{L}{\pi}\right)^2 \gamma_{\text{FP}} = \frac{1}{4}\frac{L^2}{(L+1)^2} + \mathcal{O}(L^{-4}) \to \frac{1}{4} \quad (L \to \infty). \qquad (3.33)$$

Thus we have developed a method by which we can exactly calculate the deterministic diffusion coefficient of a simple chaotic dynamical system. However, more importantly, instead of using the explicit expression for γ_{FP} given by Eq. (3.31), let us remind ourselves of the escape rate formula Eq. (2.51) for γ_{FP},

$$\gamma_{\text{FP}} = \gamma(\mathcal{C}_{B_4}) = \lambda(\mathcal{C}_{B_4}) - h_{\text{KS}}(\mathcal{C}_{B_4}), \qquad (3.34)$$

which more generally expresses this escape rate in terms of dynamical systems quantities. Combining this equation with the above equation (Eq. (3.32)) leads to our final result, the *escape rate formula for chaotic diffusion*,[5]

$$D(4) = \lim_{L \to \infty} \left(\frac{L}{\pi}\right)^2 [\lambda(\mathcal{C}_{B_4}) - h_{\text{KS}}(\mathcal{C}_{B_4})]. \qquad (3.35)$$

We have thus established a fundamental link between quantities assessing the chaotic properties of dynamical systems and the statistical physical property of diffusion.

4.   Exercises and Solutions

4.1.   *Exercises*

(1) Prove Eq. (2.7).
(2) Consider the map defined by the function

$$E(x) := \begin{cases} 2x + 1, & -1 \leq x < -1/2, \\ 2x, & -1/2 \leq x < 1/2, \\ 2x - 1, & 1/2 \leq x \leq 1. \end{cases} \qquad (4.1)$$

Draw the graph of this map. Is E ergodic? Prove your answer.
(3) Consider the Frobenius–Perron equation

$$\rho_{n+1}(x) = \sum_{x=G(x^i)} \rho_n(x^i)|G'(x^i)|^{-1} =: P\rho_n(x) \qquad (4.2)$$

for a map G defined on the real line, where $\rho_n(x)$ are probability densities at time step $n \in \mathbb{N}_0$ and P defines the Frobenius–Perron operator.

(a) Show that P is a linear and positive operator.
(b) Construct P for the map G defined in the figure below and verify that

$$\rho^*(x) = \begin{cases} 4/3, & 0 \leq x < 1/2, \\ 2/3, & 1/2 \leq x \leq 1 \end{cases} \qquad (4.3)$$

is an invariant density of the above Frobenius–Perron equation.

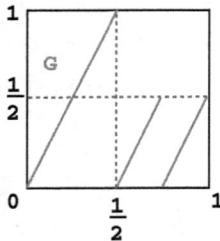

(c) By assuming that G is ergodic, calculate the Lyapunov exponent λ for this map.

(4) Consider the asymmetric tent map

$$S(x) := \begin{cases} ax, & 0 \le x < 1/a, \\ b - bx, & 1/a \le x \le 1 \end{cases} \qquad (4.4)$$

with $1/a + 1/b = 1$.

(a) Calculate the Lyapunov exponent λ for this map.
(b) Show that for the H-function defined by Eq. (2.27) in the lecture notes it holds $H(\{J_i^2\}) = 2H(\{J_i^1\})$. By assuming that $\forall n \in \mathbb{N}\, H(\{J_i^n\}) = nH(\{J_i^1\})$ calculate the KS-entropy h_{KS}. Compare your result for h_{KS} with the one obtained for λ.

(5) Consider the map $H(x) = 5x \mod 1$ on a domain which has "holes" where points escape from the unit interval. Let these escape regions be defined by the two subintervals $(0.2, 04)$ and $(0.6, 0.8)$.

(a) Sketch the map and the first two steps in the construction of its fractal repeller $\mathcal{R}_H$.
(b) Calculate the escape rate $\gamma(\mathcal{R}_H)$.
(c) Calculate the Lyapunov exponent $\lambda(\mathcal{R}_H)$
(d) Calculate the KS-entropy $h_{KS}(\mathcal{R}_H)$.
(e) By using these results, verify the escape rate formula for this map.

(6) Verify Eq. (3.33) by using Eqs. (3.32) and (3.31).

4.2. Solutions

(1) Applying the chain rule we get

$$(F^n)'(x) = (F(F^{n-1}))'(x) = F'(F^{n-1}(x))(F^{n-1})'(x)$$
$$= \cdots = F'(x_{n-1})F'(x_{n-2})\ldots F'(x_0) \qquad (4.5)$$

with $x = x_0$.

(2) We leave the drawing of the map to the reader.

Choose for g the indicator function $g : [-1, 1] \to \{0, 1\}$ with

$$g(x) := \begin{cases} 0, & -1 \leq x < 0, \\ 1, & 0 \leq x \leq 1. \end{cases} \tag{4.6}$$

We need to check Definition 2.8. For g we have

$$\int_{-1}^{1} d\mu^* \, |g(x)| = \int_{0}^{1} d\mu^* = \int_{0}^{1} dx \, \rho^*(x) < \infty, \tag{4.7}$$

because μ^* is a probability measure. Hence the assumption is fulfilled, but for $-1 \leq x < 0$ we have $\overline{g(x)} = 0$ while for $0 \leq x \leq 1$, $\overline{g(x)} = 1$. Hence $\overline{g(x)}$ depends on x, consequently $\overline{g(x)} \neq const.$ in contradiction to Definition 2.8, which implies that the map E is not ergodic.

(3) (a) Linearity of an operator P is defined by

$$P(\alpha_1 \rho^1 + \alpha_2 \rho^2) = \alpha_1 P \rho^1 + \alpha_2 P \rho^2 \tag{4.8}$$

with $\alpha_1, \alpha_2 \in \mathbb{R}$ and probability densities ρ^1, ρ^2. Positivity means $P\rho(x) \geq 0$. The proofs of these two properties for the Frobenius–Perron operator defined by Eq. (4.2) is then straightforward.

(b) From the figure we can infer

$$G(x) := \begin{cases} 2x, & 0 \leq x < 1/2, \\ 2x - 1, & 1/2 \leq x < 3/4, \\ 2x - 3/2, & 3/4 \leq x \leq 1 \end{cases} \tag{4.9}$$

and $G'(x) = 2 \, \forall x \, (x \neq 1/2, 3/4)$. We construct the piecewise inverse functions

$$x^1 = G^{-1}(x) = x/2, \qquad 0 \leq x^1 < 1/2, \, 0 \leq x < 1,$$
$$x^2 = G^{-1}(x) = (x+1)/2, \qquad 1/2 \leq x^2 < 3/4, \, 0 \leq x < 1/2, \tag{4.10}$$
$$x^3 = G^{-1}(x) = (x+3/2)/2, \quad 3/4 \leq x^1 < 1, \, 0 \leq x \leq 1/2.$$

Plugging these results into Eq. (4.2) yields

$$P\rho(x) = 1/2\rho(x/2) + 1/2\rho((x+1)/2) + 1/2\rho(x/2 + 3/4) . \tag{4.11}$$

Feeding $\rho^*(x)$ given by Eq. (4.3) into this equation shows $P\rho^*(x) = \rho^*(x)$.

c The calculation is analogous to the one of Example 2.9 yielding $\lambda = \ln 2$.

(4) (a) See again Example 2.9; in this case the solution is $\lambda = 1/a \ln a + 1/b \ln b$.

(b) The calculation follows Section 2.3:

(i) It is convenient to choose as a partition the one generated by the backward iteration of the critical point at $x_c = 1/a$.

(ii)
$$H(\{J_i^1\}) = 1/a \ln a + 1/b \ln b,$$

$$H(\{J_i^2\}) = 1/a^2 \ln a^2 + 1/(ab) \ln(ab) + 1/b^2 \ln b^2$$
$$+ 1/(ba) \ln(ba) \tag{4.12}$$
$$= 2H(\{J_i^1\}).$$

(iii) Using the stated assumption we get

$$h(\{J_i^n\}) = \lim_{n\to\infty} \frac{1}{n} H(\{J_i^n\}) = H(\{J_i^1\}). \tag{4.13}$$

(iv) Since the partition above is generating we find

$$h_{\mathrm{KS}} = h(\{J_i^n\}) = 1/a \ln a + 1/b \ln b = \lambda \tag{4.14}$$

according to Pesin's theorem.

(5) (a) The sketch can be performed in analogy to Fig. 7 and is left to the reader.

(b) Since $\rho^*(x) = 1$, just consider the Lebesgue measure of the sets $\{R_i^n\}$, i.e. the total lengths l_n, which are $l_0 = 1$, $l_1 = 3/5$, $l_2 = 9/25$. It follows that $l_n = (3/5)^n = \exp(-n \ln(5/3))$, hence $\gamma(\mathcal{R}_H) = \ln(5/3)$.

(c) It is easy to find $\lambda(\mathcal{R}_H) = \ln 5$.

(d) The calculation is in analogy to Example 2.23 and yields $h_{\mathrm{KS}}(\mathcal{R}_H) = \ln 3$.

(e) We thus have the escape rate formula $\gamma(\mathcal{R}_H) = \ln(5/3) = \lambda(\mathcal{R}_H) - h_{\mathrm{KS}}(\mathcal{R}_H)$.

(6) With $\cos x \simeq 1 - x^2/2$ and $\ln(1 \pm x) \simeq \pm x$ we have

$$\gamma_{\text{FP}} \simeq \ln \frac{4}{2 + 2 - (\pi/(L+1))^2} \simeq \frac{1}{4}\left(\frac{\pi}{L+1}\right)^2, \qquad (4.15)$$

which leads to Eq. (3.33).

References

1. H.G. Schuster, *Deterministic Chaos: An Introduction*, 2nd edn. VCH, Weinheim (1989).
2. E. Ott, *Chaos in Dynamical Systems*. Cambridge University Press, Cambridge (1993).
3. C. Beck and F. Schlögl, *Thermodynamics of Chaotic Systems*. Cambridge Nonlinear Science Series, Vol. 4, Cambridge University Press, Cambridge (1993).
4. J.R. Dorfman, *An Introduction to Chaos in Nonequilibrium Statistical Mechanics*. Cambridge University Press, Cambridge (1999).
5. P. Gaspard, *Chaos, Scattering, and Statistical Mechanics*. Cambridge University Press, Cambridge (1998).
6. R. Klages, *Microscopic Chaos, Fractals and Transport in Nonequilibrium Statistical Mechanics*. Advanced Series in Nonlinear Dynamics, Vol. 24, World Scientific, Singapore (2007).
7. K.T. Alligood, T.S. Sauer and J.A. Yorke, *Chaos — An Introduction to Dynamical Systems*. Springer, New York (1997).
8. V.I. Arnold and A. Avez, *Ergodic Problems of Classical Mechanics*. W.A. Benjamin, New York (1968).
9. A. Lasota and M.C. Mackay, *Chaos, Fractals, and Noise*. Springer (1994).
10. A. Katok and B. Hasselblatt, *Introduction to the Modern Theory of Dynamical Systems*. Encyclopedia of Mathematics and its Applications, Vol. 54, Cambridge University Press, Cambridge (1995).
11. F. Reif, *Fundamentals of Statistical and Thermal Physics*. McGraw-Hill, Auckland (1965).

Chapter 2

Aperiodic Order

Uwe Grimm

School of Mathematics and Statistics, The Open University
Walton Hall, Milton Keynes MK7 6AA, UK
uwe.grimm@open.ac.uk

Order phenomena are ubiquitous in nature. Yet there is no obvious way to define what precisely constitutes order, nor do we know what manifestations of order may exist. This chapter presents a brief introduction to the theory of aperiodic order, concentrating on sequences and tilings.

1. Introduction

Crystals are the paradigm of order in the natural world. Their perfect facets and symmetries are a visible consequence of the underlying order of the arrangements of atoms in the crystal. The conventional assumption that this arrangement is periodic in space implies that the structure of a crystal is described by a lattice. This limits the possible symmetries to those compatible with lattice periodicity, and a complete classification of crystal structures has been obtained over a century ago.

However, in 1982 Dan Shechtman[1] discovered a new type of crystalline substance that does not conform to this classification, and displays a crystallographically 'forbidden' symmetry. His discovery of quasicrystals was honoured by the award of the 2011 Nobel Prize in Chemistry, and has revolutionised crystallography. Crystals now include classes of non-periodically ordered structures, and today we know many materials that can generate this intricate type of order.

The discovery of quasicrystals inspired research activity into aperiodically ordered structures, in particular concerning tilings of space, which can be thought of as generalisations of lattices that can serve as structure models for quasicrystals. However, the mathematics of aperiodic order predates the discovery of quasicrystals. In the 1960s, sets of tiles that can tile the plane but do not admit any periodic tiling were of interest in connection with decidability questions,[2] and Penrose published his celebrated tiling in 1974.[3] Over the past decades, aperiodic order has developed into a thriving field with many connections to other areas of mathematics. An example of an aperiodic tiling is shown in Fig. 1.

This chapter can only offer a glimpse into the world of aperiodic order. For further introductory material we refer to Refs. 5–7. Generally, we will state results without proofs, but give explicit references to the literature. Most of these refer to Ref. 8, which is also our general reference for more on the background and for further details. We start with a collection of definitions of various notions for point sets in Euclidean space.

2.   Point Sets and Lattices

We are considering subsets of d-dimensional Euclidean space $\mathbb{R}^d$. A set consisting of one point in $\mathbb{R}^d$ is called a *singleton set*, and countable unions of singleton sets are called *point sets*.

A point set $\Lambda \subset \mathbb{R}^d$ is *discrete* if each element $x \in \Lambda$ has an open neighbourhood $U = U(x) \subset \mathbb{R}^d$ that does not contain any other point of Λ. For each $x \in \Lambda$, there is then a radius $r > 0$ such that $B_r(x)$ (the open ball of radius r around x) satisfies $B_r(x) \cap \Lambda = \{x\}$. The point set Λ is called *uniformly discrete* if there is an open neighbourhood U of $0 \in \mathbb{R}^d$ such that $(x + U) \cap (y + U) = \varnothing$ holds for *all* distinct $x, y \in \Lambda$. Here, $x+U := \{x+u \mid u \in U\}$ and, more generally, we define the *Minkowski sum* and *difference* of two arbitrary sets $U, V \subset \mathbb{R}^d$ as $U \pm V := \{u \pm v \mid u \in U, v \in V\}$. Clearly, uniform discreteness is a stronger property than discreteness, and implies that there exists an $r > 0$ such that the distance between any two points in the set satisfies $|x - y| > r$.

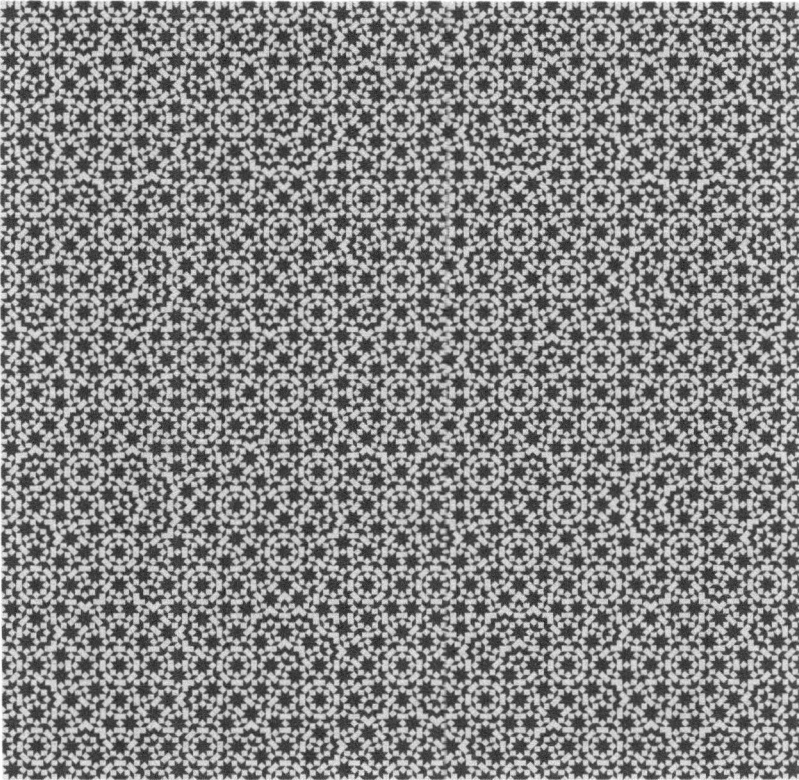

Fig. 1.   A large patch of an eightfold symmetric tiling by Watanabe, Soma and Ito.[4] It highlights the long-range (aperiodic) order in the structure. The tiling actually consists of two types of tiles, squares (light grey) and 45-degree rhombuses (dark grey), which may combine to form different shapes such as rectangles and stars. The construction and properties of such tilings will be discussed below.

A point set $\Lambda \subset \mathbb{R}^d$ is called *locally finite* if, for all compact $K \subset \mathbb{R}^d$, the intersection $K \cap \Lambda$ is a finite set (or empty). A point set $\Lambda \subset \mathbb{R}^d$ is *relatively dense* if a compact $K \subset \mathbb{R}^d$ exists such that $\Lambda + K = \mathbb{R}^d$. This property means that Λ does not contain arbitrarily large 'holes', in the sense that there exists a radius $R > 0$ such that $B_R(x) \cap \Lambda \neq \emptyset$ for any $x \in \mathbb{R}^d$.

A *cluster* of a point set $\Lambda \subset \mathbb{R}^d$ is the intersection $K \cap \Lambda$ for some compact $K \subset \mathbb{R}^d$. A point set $\Lambda \subset \mathbb{R}^d$ has *finite local complexity*

(FLC) with respect to translations when the collection $\{(t+K)\cap\Lambda \mid t \in \mathbb{R}^d\}$, for any given compact $K \subset \mathbb{R}^d$, contains only finitely many clusters up to translations. This means that the set has only finitely many 'motifs' in which points can be arranged locally around a given point of the set.

A point set $\Lambda \subset \mathbb{R}^d$ that is both uniformly discrete and relatively dense is called a *Delone set* (or *Delaunay set*). An even stronger property is the following. A point set $\Lambda \subset \mathbb{R}^d$ is a *Meyer set*, if Λ is relatively dense and the difference set $\Lambda - \Lambda$ is uniformly discrete. The various properties introduced above are related as follows:

$$\Lambda \text{ Meyer} \Longrightarrow \Lambda \text{ FLC and Delone} \Longrightarrow \Lambda \text{ Delone.}$$

The following lemma is a useful characterisation of Meyer sets.

Lemma 2.1 (Lemma 2.1 in Ref. 8). *Let $\Lambda \subset \mathbb{R}^d$ be a Delone set, such that $\Lambda - \Lambda \subset \Lambda + F$ for some finite set $F \subset \mathbb{R}^d$. Then Λ is a Meyer set.*

In Theorem 3.1 of Ref. 9, Lagarias showed that the converse is true in $\mathbb{R}^d$, and even more generally.

Exercise 2.2. For each of the point sets in $\mathbb{R}$ below, decide whether they are relatively dense, uniformly discrete, Delone, FLC or Meyer. Here, $\mathbb{N} = \{n \in \mathbb{Z} \mid n > 0\}$ denotes the natural numbers:

(a) $\Lambda_a = 2\,\mathbb{Z}$;
(b) $\Lambda_b = \{n + 1/n \mid n \in \mathbb{Z}\backslash\{0\}\}$;
(c) $\Lambda_c = -\mathbb{N} \cup \{0\} \cup \sqrt{3}\,\mathbb{N}$;
(d) $\Lambda_d = \mathbb{Z}\backslash S$, where S is an arbitrary subset of $2\,\mathbb{Z}$.

2.1. *Lattices and periodicity*

A point set $\Gamma \subset \mathbb{R}^d$ is called a *lattice* in $\mathbb{R}^d$ if there exist d vectors $b_1, \ldots, b_d$ such that

$$\Gamma = \mathbb{Z}b_1 \oplus \cdots \oplus \mathbb{Z}b_d := \left\{ \sum_{i=1}^{d} m_i b_i \,\middle|\, \text{all } m_i \in \mathbb{Z} \right\}, \qquad (2.1)$$

together with the requirement that its $\mathbb{R}$-span $\langle \Gamma \rangle_{\mathbb{R}} = \mathbb{R}^d$. The latter requirement means that the vectors are linearly independent and form an $\mathbb{R}$-basis of $\mathbb{R}^d$. The set $\{b_1, \ldots, b_d\}$ is then called a *basis* of the lattice Γ. Here, of course, it is understood to be a $\mathbb{Z}$-basis, so all integer linear combinations of the vectors b_i produce the lattice Γ, while all real linear combinations exhaust the entire space $\mathbb{R}^d$. Clearly, any lattice $\Gamma \subset \mathbb{R}^d$ is a Meyer set and, consequently, a Delone set of finite local complexity.

Let Λ be a point set in $\mathbb{R}^d$. An element $t \in \mathbb{R}^d$ is called a *period* of Λ when $t + \Lambda = \Lambda$. The set $\mathrm{per}(\Lambda) := \{t \in \mathbb{R}^d \mid t + \Lambda = \Lambda\}$, called the *set of periods* of Λ, is a subgroup of $\mathbb{R}^d$.

A point set $\Lambda \subset \mathbb{R}^d$ is called *periodic* (of *rank m*) when $\mathrm{per}(\Lambda) \subset \mathbb{R}^d$ is non-trivial (with $1 \leq m = \dim\langle \mathrm{per}(\Lambda) \rangle_{\mathbb{R}} \leq d$), and *non-periodic* when $\mathrm{per}(\Lambda) = \{0\}$. The point set Λ is called *crystallographic* when $\mathrm{per}(\Lambda)$ is a lattice in $\mathbb{R}^d$, and *non-crystallographic* otherwise. Crystallographic point sets are characterised as follows.

Proposition 2.3 (Proposition 3.1 in Ref. 8). *A locally finite point set $\Lambda \subset \mathbb{R}^d$ is crystallographic if and only if there is a lattice $\Gamma \subset \mathbb{R}^d$ and a finite point set $F \subset \mathbb{R}^d$ such that $\Lambda = \Gamma + F$.*

Such a point set $\Lambda = \Gamma + F \subset \mathbb{R}^d$ is called a *crystallographic point packing* in $\mathbb{R}^d$. Another interpretation is that the packing Λ consists of a finite union of translates of the lattice Γ.

2.2. *Crystallographic restriction*

Clearly, a rotated lattice is again a lattice, but only certain rotations will leave a lattice invariant. Such rotations are of interest for crystallography, as they correspond to symmetries of the crystal. It turns out that compatibility of rotations and translations leaves only very limited possibilities, as summarised in the following classic results, which are known as the crystallographic restriction. Because these are fundamental results in crystallography, we will include brief proofs.

Lemma 2.4 (Lemma 3.2 in Ref. 8). *Consider a lattice $\Gamma \subset \mathbb{R}^d$. If $R\Gamma = \Gamma$ for $R \in \mathrm{O}(d)$, the corresponding characteristic polynomial $P(\lambda) = \det(R - \lambda \mathbf{1})$ has integer coefficients only.*

Proof. The claim is a consequence of the fact that lattice points have to be mapped onto lattice points, which implies that $R = BAB^{-1}$ where B is the matrix of basis vectors of the lattice Γ and A is an integer matrix. Hence R and A share the same characteristic polynomial, which then has integer coefficients only. ☐

Corollary 2.5 (Corollary 3.1 in Ref. 8). *A lattice $\Gamma \subset \mathbb{R}^d$ with $d = 2$ or $d = 3$ can have n-fold rotational symmetry at most for $n \in \{1, 2, 3, 4, 6\}$.*

Proof. For $d = 2$, the rotation matrix for a rotation by an angle φ,

$$R_\varphi = \begin{pmatrix} \cos(\varphi) & -\sin(\varphi) \\ \sin(\varphi) & \cos(\varphi) \end{pmatrix},$$

has characteristic polynomial $P_2(\lambda) = \lambda^2 - 2\cos(\varphi) + 1$. Lemma 2.4 stipulates that $2\cos(\varphi) \in \mathbb{Z}$, hence $|\cos(\varphi)| \in \{0, \frac{1}{2}, 1\}$, which gives the desired result. For $d = 3$, the result follows from Euler's rotation theorem. Choosing coordinates such that one direction is along the fixed rotation axis, it is clear that $P_3(\lambda) = (1 - \lambda)P_2(\lambda)$ for a suitable rotation angle, and hence the same argument applies. ☐

Note that similar restrictions apply in higher dimensions, but new rotational symmetries will become available. In order to observe a rotational symmetry of order n, the lattice has to have a minimum dimension d_n which is given by a function Φ which is defined by $\Phi(p^r) = \phi(p^r) = p^{r-1}(p - 1)$ for any powers of a prime p with $r \geq 1$, where ϕ is Euler's totient function

$$\phi(n) := \operatorname{card}\{1 \leq k \leq n \mid \gcd(k, n) = 1\}. \tag{2.2}$$

For the remaining values of n, it is defined by the relation $\Phi(2n) = \Phi(n)$ for all odd n and by $\Phi(mn) = \Phi(m) + \Phi(n)$ for coprime integers m and n (excluding $m = 2$ and $n = 2$); see Theorem 3.1 in Ref. 8 for details.

2.3. *Cyclotomic fields and algebraic integers*

To describe non-crystallographic symmetries in the plane, it is convenient to identify $\mathbb{R}^2$ with the complex plane $\mathbb{C}$, which means that a rotation corresponds to multiplication by a complex number of unit modulus. We now make a brief excursion to number theory and discuss some elementary properties of cyclotomic fields.

Let $\xi_n \in \mathbb{C}$ be a primitive nth root of unity (with $n > 2$), so that $\xi_n^m = 1$ precisely when $n|m$. For example, you can choose $\xi_n = \exp(2\pi i/n)$. The *cyclotomic field* $\mathbb{Q}(\xi_n)$ is a field extension of $\mathbb{Q}$ of degree $\phi(n)$; the elements of $\mathbb{Q}(\xi_n)$ are $\mathbb{Q}$-linear combinations of powers of ξ_n. Within this number field, the set $\mathbb{Z}[\xi_n]$ is the ring of integers, comprising the $\mathbb{Z}$-linear combinations of powers of ξ_n. The maximal real subfield of $\mathbb{Q}(\xi_n)$ is given by $\mathbb{Q}(\xi_n + \bar{\xi}_n)$, with relative degree 2 (for $n > 2$). Its ring of integers is $\mathbb{Z}[\xi_n + \bar{\xi}_n]$.

Exercise 2.6. Consider $\mathbb{Z}[\xi] = \{a_0 + a_1\xi + a_2\xi^2 + a_3\xi^3 \mid a_0, a_1, a_2, a_3 \in \mathbb{Z}\}$, where $\xi = \exp(2\pi i/5)$ is a primitive fifth root of unity. Show that $\mathbb{Z}[\xi]$, when interpreted as a point set in $\mathbb{R}^2$, is invariant under rotations by $\pi/5$.

Certain algebraic numbers will play a central role below, so we briefly introduce them here. Remember that an *algebraic number* is a (real or complex) root of a polynomial with integer coefficients (and hence all roots of unity are algebraic numbers, as they are roots of $x^n - 1$ for some n). If the leading coefficient is 1, the roots are called *algebraic integers*. If in addition the constant term is ± 1, they are called *algebraic units*.

A real algebraic integer $\alpha > 1$ is called a *Pisot–Vijayaraghavan number*, or *PV number* for short, if all its algebraic conjugates (apart from α itself) lie inside the open unit disk.

Example 2.7. The golden ratio $\tau = (1 + \sqrt{5})/2 \approx 1.618$ is an algebraic integer (in fact an algebraic unit) of degree 2, as a root of $x^2 - x - 1 = 0$. Its algebraic conjugate is $\tau' = (1 - \sqrt{5})/2 = 1 - \tau \approx -0.618$, so τ is a PV number (in fact, a PV unit).

A real algebraic integer $\alpha > 1$ is called a *Salem number*, if all its algebraic conjugates (apart from α itself) lie inside the closed unit disk, with at least one conjugate on the unit circle. The importance of Pisot–Vijayaraghavan and Salem numbers in our context is apparent from the following important theorem by Lagarias.[9]

Theorem 2.8 (Theorem 4.1 in Ref. 9). *If $\Lambda \subset \mathbb{R}^d$ is a Meyer set with $\alpha\Lambda \subset \Lambda$ for some $\alpha > 1$, then α is a PV or a Salem number.*

Exercise 2.9. Let $\tau = (1 + \sqrt{5})/2$ be the golden ratio. Show that the ring of integers $\mathbb{Z}[\tau] = \{m + n\tau \mid m, n \in \mathbb{Z}\}$ satisfies $\tau\mathbb{Z}[\tau] = \mathbb{Z}[\tau]$.

2.4. *Minkowski embedding*

Here, we concentrate on the example of $\mathbb{Z}[\tau] = \{m + n\tau \mid m, n \in \mathbb{Z}\}$ to motivate a natural embedding of a ring of algebraic integers into a lattice. The embedding uses algebraic conjugation $x \mapsto x'$ in the corresponding field, in this case $\mathbb{Q}(\sqrt{5})$. Here, algebraic conjugation is simply defined by the map $\sqrt{5} \mapsto -\sqrt{5}$ and its extension to a field automorphism. The *diagonal* or *Minkowski embedding*

$$\mathcal{L} = \big\{(x, x') \mid x \in \mathbb{Z}[\tau]\big\} \tag{2.3}$$

defines a lattice in $\mathbb{R}^2$. From $(m+n\tau, m+n\tau') = m(1,1)+n(\tau,\tau')$ for all $m, n \in \mathbb{Z}$, it is clear that $\mathcal{L}$ is the lattice generated by the vectors $(1, 1)$ and (τ, τ'). This embedding is sketched in Fig. 2.

The Minkowski embedding of real algebraic integers of rank m into $\mathbb{R}^m$ is defined analogously in terms of its algebraic conjugates (the other roots of the irreducible polynomial defining the algebraic integer); see Ref. 8 for details.

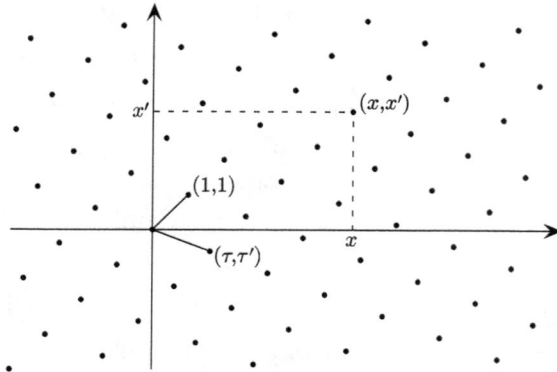

Fig. 2. Embedding of the numbers in $\mathbb{Z}[\tau]$ into a two-dimensional lattice.

Exercise 2.10. Consider the ring $\mathbb{Z}[\sqrt{2}] = \{m + n\sqrt{2} \mid m, n \in \mathbb{Z}\}$. Using the algebraic conjugation $\sqrt{2} \longmapsto -\sqrt{2}$ in $\mathbb{Q}(\sqrt{2})$, construct the Minkowski embedding of $\mathbb{Z}[\sqrt{2}]$.

3. Symbolic Dynamics

Many example of aperiodic structures arise from substitution rules. Therefore, we now embark on a brief excursion into the field of symbolic dynamics.

3.1. *Substitution sequences*

Consider n-letter alphabet $\mathcal{A}_n = \{a_i \mid 1 \leq i \leq n\}$ of letters a_i, and the free group $F_n := \langle a_1, \ldots, a_n \rangle$ generated by the letters (and their formal inverses). A *general substitution rule* ϱ on an n-letter alphabet $\mathcal{A}_n$ is an endomorphism of the corresponding free group F_n. This means that $\varrho(uv) = \varrho(u)\varrho(v)$ and $\varrho(u^{-1}) = \left(\varrho(u)\right)^{-1}$. While this general setting has some advantages, we will limit ourselves to substitution rules ϱ where the images $\varrho(a_i)$ of the letters contain no negative powers of the letters. We associate to a substitution ϱ a *substitution matrix* $M(\varrho) \in \mathrm{Mat}(n, \mathbb{Z})$, which is defined by

$$\left(M(\varrho)\right)_{i,j} = \mathrm{card}_{a_i}\left(\varrho(a_j)\right).$$

Its entries count how often each letter appears in the image of the letters under ϱ. Let us consider a central example.

Example 3.1 (Fibonacci substitution). The Fibonacci substitution is defined by the two-letter substitution rule

$$\varrho : \begin{array}{l} a \longmapsto ab \\ b \longmapsto a \end{array} \quad \text{with substitution matrix} \quad M_\varrho = \begin{pmatrix} 1 & 1 \\ 1 & 0 \end{pmatrix}. \quad (3.1)$$

Iterating ϱ on an initial word $w^{(0)} = a$ gives

$$a \longmapsto ab \longmapsto aba \longmapsto abaab \longmapsto abaababa \longmapsto \ldots \longmapsto w^{(m)} \xrightarrow{m \to \infty} w \quad (3.2)$$

which converges (in an appropriate topology) to a one-sided fixed point $w = \varrho(w)$. These words are related to the celebrated Fibonacci numbers f_m as follows:

$$|w^{(m)}| = f_{m+2} \quad \text{with } \text{card}_a(w^{(m)}) = f_{m+1} \text{ and } \text{card}_b(w^{(m)}) = f_m, \tag{3.3}$$

where $f_0 = 0$, $f_1 = 1$ and $f_{m+1} = f_m + f_{m-1}$. Note that this definition extends to negative values of m. In the limit as $m \to \pm\infty$, the ratios of successive Fibonacci numbers approach the golden ratio or its algebraic conjugate

$$\lim_{m \to \pm\infty} \frac{f_{m+1}}{f_m} = \frac{1 \pm \sqrt{5}}{2} = \begin{cases} \tau, \\ \tau'. \end{cases}$$

Alternatively, the words $w^{(m)} = \varrho^m(w^{(0)})$ can be defined by the recursion $w^{(m+1)} = w^{(m)}w^{(m-1)}$, which follows from the substitution rule of Eq. (3.1).

Two-sided Fibonacci sequences can be defined by considering the action of ϱ on a two-letter seed, with the vertical bar denoting the location of the origin,

$$a|a \xmapsto{\varrho} \underline{ab}|ab \xmapsto{\varrho} a\underline{ba}|aba$$

$$\xmapsto{\varrho} aba\underline{ab}|abaab \xmapsto{\varrho} abaab\underline{aba}|abaababa$$

$$\xmapsto{\varrho} abaababaaba\underline{ab}|abaababaabaab \xmapsto{\varrho} \cdots$$

As indicated, this results in a limiting 2-cycle, which implies that there are two fixed points under ϱ^2.

A (non-negative) substitution rule ϱ on a finite alphabet $\mathcal{A}_n$ is called *primitive* when some $k \in \mathbb{N}$ exists such that every a_j is a subword of each $\varrho^k(a_i)$. A substitution rule ϱ is primitive precisely when M_ϱ is a primitive non-negative integer matrix.

The substitution matrix determines the statistics of letters in the substitution sequence. Under m-fold substitution, we find

$$\begin{pmatrix} \text{card}_a(\varrho^m(u)) \\ \text{card}_b(\varrho^m(u)) \end{pmatrix} = M_\varrho^m \begin{pmatrix} \text{card}_a(u) \\ \text{card}_b(u) \end{pmatrix},$$

and it follows that the right eigenvector for the leading (Perron–Frobenius) eigenvalue of M_ϱ encodes the letter frequencies in the limit as $m \to \infty$.

Exercise 3.2. Consider the inflation rule

$$\varrho : \begin{aligned} a &\mapsto abb \\ b &\mapsto a \end{aligned}$$

on the two-letter alphabet $\{a, b\}$.

(a) Show that ϱ is primitive.
(b) Calculate the corresponding substitution matrix M and its leading eigenvalue λ.
(c) If you start from the initial word $w^{(0)} = a$ and define $w^{(n+1)} = \varrho(w^{(n)})$ for $n \geq 1$, show that $w^{(n+1)} = w^{(n)} w^{(n-1)} w^{(n-1)}$.
(d) How many letters of each type does the word $w^{(8)}$ have?
(e) Calculate the right eigenvector of M to the eigenvalue λ. What is the frequency of letters a and b in a fixed point word? Compare this with the result for $w^{(8)}$ obtained above.

A finite word is called *legal* for a substitution rule ϱ on a finite alphabet if it occurs as a subword of $\varrho^k(a_i)$ for some $1 \leq i \leq n$ and some $k \in \mathbb{N}$. Using a legal word as a seed, bi-infinite (two-sided) substitution sequences $w = \ldots w_{-2} w_{-1} | w_0 w_1 w_2 \ldots \in \mathcal{A}_n^{\mathbb{Z}}$ are obtained as fixed points of ϱ^k for a suitable value of k. It is easy to show that such fixed points always exist. To define the convergence of such sequences, the *local topology* is employed, in which two sequences w and w' are close when they agree on a large region around index 0. Convergence of a sequence of finite words (of increasing length) is implicitly considering them as embedded objects in $\mathcal{A}_n^{\mathbb{Z}}$.

Exercise 3.3. Consider the substitution rule ϱ of Exercise 3.2.

(a) Consider the four possible two-letter seeds $a|a$, $a|b$, $b|a$ and $b|b$. Which of these are legal?
(b) Iterate ϱ on the legal seeds. Are there any fixed points under ϱ?
(c) Find all fixed points under ϱ^2.

3.2. *The hull and local indistinguishability*

In symbolic dynamics, one considers the action of the *shift operator* S on $\mathcal{A}_n^{\mathbb{Z}}$. It simply acts on a word $w \in \mathcal{A}_n^{\mathbb{Z}}$ by shifting the letters by one position according to $(Sw)_i := w_{i+1}$. An S-invariant closed subset $X \subset \mathcal{A}_n^{\mathbb{Z}}$ is called a two-sided *shift space*.

Given a sequence $w \in \mathcal{A}_n^{\mathbb{Z}}$, the shift space $\mathbb{X}(w) := \overline{\{S^i w \mid i \in \mathbb{Z}\}}$ is called the (two-sided, symbolic or discrete) *hull* of w, where the line denotes the closure in the local topology. The hull of a substitution ϱ is defined as the hull of a fixed point of a suitable power ϱ^k. The hull, together with the $\mathbb{Z}$-action of the shift operator, defines a *topological dynamical system* $(\mathbb{X}(w), \mathbb{Z})$. Here, we have a continuous $\mathbb{Z}$-action of the shift on the compact space $\mathbb{X}(w)$, and the additional action of the substitution ϱ on $\mathbb{X}(w)$, which is continuous as well. A two-sided shift space $\mathbb{X} \subset \mathcal{A}^{\mathbb{Z}}$ is called *minimal* when, for all $w \in \mathbb{X}$, the shift orbit $\{S^i w \mid i \in \mathbb{Z}\}$ is dense in $\mathbb{X}$.

Two (bi-infinite) words u and v in the same alphabet are *locally indistinguishable* (LI), denoted by $u \overset{\text{LI}}{\sim} v$, when each finite subword of u is also a subword of v and vice versa. The *LI class* of a word $w \in \mathcal{A}^{\mathbb{Z}}$ is $\mathrm{LI}(w) := \{z \in \mathcal{A}^{\mathbb{Z}} \mid z \overset{\text{LI}}{\sim} w\}$. The following results highlight the connection between LI classes and the hull.

Lemma 3.4 (Lemma 4.2 in Ref. 8). *If w is a bi-infinite word, its LI class is contained in the hull of w, and one has $\mathbb{X}(w) = \overline{\mathrm{LI}(w)}$. In particular, $\mathbb{X}(u) = \mathbb{X}(v)$ holds for any two bi-infinite words $u \overset{LI}{\sim} v$.*

Proposition 3.5 (Proposition 4.1 in Ref. 8). *If w is a bi-infinite word in the finite alphabet $\mathcal{A}$, with LI class $\mathrm{LI}(w)$ and hull $\mathbb{X}(w)$, the following assertions are equivalent:*

(1) $\mathbb{X}(w)$ *is minimal;*
(2) $\mathrm{LI}(w)$ *is closed;*
(3) $\mathbb{X}(w) = \mathrm{LI}(w)$.

The final result of this part shows that it does not matter which fixed point of a substitution is used to define its hull.

Proposition 3.6 (Proposition 4.2 in Ref. 8). *Let ϱ be a primitive substitution rule on a finite alphabet. Then, any two bi-infinite* *fixed points u and v of ϱ are LI. The same conclusion holds if u and v are fixed points of possibly different positive powers of ϱ.*

3.3. *Repetitivity and aperiodicity*

We are now looking at generalisations of the notion of periodicity. A bi-infinite word w (over a finite alphabet) is called *repetitive* when every finite subword of w reappears in w with bounded gaps. Repetitivity of w is linked to minimality of the hull by the following result.

Proposition 3.7 (Proposition 4.3 in Ref. 8). *If w is a bi-infinite word in a finite alphabet, the hull $\mathbb{X}(w)$ is minimal if and only if w is repetitive.*

Primitive substitutions are a particularly nice class due to the following properties.

Lemma 3.8 (Lemma 4.4 in Ref. 8). *Any bi-infinite fixed point of a primitive substitution on a finite alphabet is repetitive.*

Theorem 3.9 (Theorem 4.1 in Ref. 8). *Every primitive substitution rule on a finite alphabet possesses a unique hull. This hull consists of a single, closed LI class.*

Clearly, a bi-infinite sequence w is periodic if there exists a $k > 0$ such that $S^k w = w$, and non-periodic if no such k exists. We define a stronger property as follows. A bi-infinite sequence w in a finite alphabet is called (topologically) *aperiodic* when $\mathbb{X}(w)$ contains no periodic sequence. A primitive substitution rule ϱ is *aperiodic* when its unique hull contains no periodic element. This definition of aperiodicity rules out 'trivial' cases of non-periodicity, such as a sequence consisting of letters a everywhere apart from a single letter b at one position. Clearly, such a sequence cannot be periodic, but the hull will contain the periodic sequence consisting just of letters a, as a limit of a sequence where the position of the letter b is moving off to infinity. So this example is non-periodic but not aperiodic.

For repetitive words, the hull is either finite (which happens when the word is periodic) or it is a Cantor set.

Proposition 3.10 (Proposition 4.5 in Ref. 8). *The symbolic hull of a repetitive, bi-infinite word over a finite alphabet is either finite or a Cantor set.*

While proving aperiodicity of a sequence may be difficult in general, primitive substitution rules provide plenty of examples.

Theorem 3.11 (Theorem 4.6 in Ref. 8). *Let ϱ be a primitive substitution rule on a finite alphabet with substitution matrix M_ϱ, and let w be a bi-infinite fixed point of ϱ. If the Perron–Frobenius eigenvalue of M_ϱ is irrational, the sequence w is aperiodic.*

3.4. *Geometric inflation*

Finally, we would like to link the symbolic context of substitutions on a finite alphabet with the geometric concept of a tiling, in this case a tiling of the real line by intervals. For a primitive substitution ϱ on a finite alphabet with substitution matrix M_ϱ and Perron–Frobenius eigenvalue λ, the associated *geometric inflation rule* with inflation multiplier λ is obtained by turning the letters a_i into closed intervals (the *prototiles*) with lengths proportional to the entries of the left Perron–Frobenius eigenvector of M_ϱ, and by dissecting the λ-inflated prototiles into copies of the original ones, respecting the order specified by ϱ. Let us explain this by means of our favourite example.

Example 3.12 (Fibonacci inflation). For the Fibonacci substitution of Eq. (3.1), we obtain the leading eigenvalue $\lambda = \tau = \frac{1}{2}(1+\sqrt{5})$. As the substitution matrix is symmetric, both left and right eigenvectors for this eigenvalue are proportional to $(\tau, 1)$. The right eigenvector implies that the frequency of the letter a is τ times the frequency of the letter b (in the infinite limit word), while the left eigenvector shows that the appropriate geometric realisation requires two intervals (one for each letter) with a length ratio of $\tau{:}1$ (the longer interval corresponding to the more frequent letter, which in our case is the letter a). This results in the following geometric inflation rule

This rule consists of two steps, a re-scaling of the intervals by the factor λ and a subsequent dissection of the re-scaled intervals into intervals of the original size. Choosing the interval length ratio according to the left Perron–Frobenius eigenvector of the substitution matrix guarantees that the resulting inflation rule is consistent, in the sense that the combined lengths of the intervals corresponding to $\varrho(a_i)$ equals λ times the length of the interval for a_i.

Exercise 3.13. Consider again the substitution rule ϱ of Exercise 3.2.

(a) If the short interval is chosen to have length 1, what is the length of the long interval in the corresponding geometric inflation rule?
(b) Consider the point set Λ of all left interval endpoints in the infinite tiling which corresponds to fixed point word, and calculate the average distance of points. What is the density of the point set Λ?

4. Patterns and Tilings

After having made the connections between symbolic substitutions and one-dimensional tilings, we are now moving on to discuss inflations in two (or more) dimensions. We start by defining patterns and tilings, and what we mean by an inflation of a tiling.

A *pattern* $\mathcal{T}$ in Euclidean space $\mathbb{R}^d$ is a non-empty set of non-empty subsets of $\mathbb{R}^d$. We refer to the elements of $\mathcal{T}$ as the *fragments* of the pattern $\mathcal{T}$, and write $\mathcal{T} \sqsubset \mathbb{R}^d$ for a pattern $\mathcal{T}$ in $\mathbb{R}^d$.

Example 4.1. A locally finite point set $\Lambda \subset \mathbb{R}^d$ can be interpreted as a pattern

$$\mathcal{T}_\Lambda = \big\{ \{x\} \mid x \in \Lambda \big\},$$

where we tacitly identify Λ and $\mathcal{T}_\Lambda$.

A *tiling* in $\mathbb{R}^d$ is a pattern $\mathcal{T} = \{T_i \mid i \in I\} \sqsubset \mathbb{R}^d$, with (countable) index set I and non-empty closed sets $T_i \subset \mathbb{R}^d$, subject to the conditions $\bigcup_{i \in I} T_i = \mathbb{R}^d$ and $T_i^\circ \cap T_j^\circ = \varnothing$ for all $i \neq j$. The fragments T_i of $\mathcal{T}$ are called the *tiles* of the tiling, and their equivalence classes up to translations (or, alternatively, up to congruence) are called *prototiles*.

We will now generalise some of the notions introduced above in the symbolic context to patterns and tilings.

A pattern $\mathcal{T} \sqsubset \mathbb{R}^d$ is called *locally finite* if $\mathcal{T} \sqcap K$ has finite cardinality, for all compact $K \subset \mathbb{R}^d$. Here and in what follows, the intersection is defined as $\mathcal{T} \sqcap A := \{T \in \mathcal{T} \mid T \cap A \neq \varnothing\}$, which means it contains all fragments in $\mathcal{T}$ that intersect A. Let $\mathcal{T} \sqsubset \mathbb{R}^d$ be a locally finite pattern. When $K \subset \mathbb{R}^d$ is compact, the pattern $\mathcal{T} \sqcap K$ is called a *cluster* of $\mathcal{T}$. We also speak of a *patch* when K is convex.

Two (locally finite) patterns $\mathcal{T}$ and $\mathcal{T}'$ in $\mathbb{R}^d$ are *locally indistinguishable*, or LI for short and written as $\mathcal{T} \overset{\text{LI}}{\sim} \mathcal{T}'$, when any cluster of $\mathcal{T}$ occurs also in $\mathcal{T}'$ and vice versa. This means that, for any compact $K \subset \mathbb{R}^d$, there are translations $t, t' \in \mathbb{R}^d$ such that $\mathcal{T} \sqcap K = (-t' + \mathcal{T}') \sqcap K$ together with $\mathcal{T}' \sqcap K = (-t + \mathcal{T}) \sqcap K$.

Two patterns $\mathcal{T}, \mathcal{T}' \sqsubset \mathbb{R}^d$ are ε-close in the *local topology* when

$$\mathcal{T} \sqcap \overline{B_{1/\varepsilon}(0)} = (-t + \mathcal{T}') \sqcap \overline{B_{1/\varepsilon}(0)}$$

holds for some $t \in B_\varepsilon(0)$.

4.1. *Local derivability*

Local indistinguishability is an equivalence of patterns. More generally, we are interested in classifying patterns that are essentially representing the same structure, in the sense that one can be transformed into the other by local operations. The following definitions provide the appropriate concept.

A pattern $\mathcal{T}' \sqsubset \mathbb{R}^d$ is said to be *locally derivable* from a pattern $\mathcal{T} \sqsubset \mathbb{R}^d$, written as $\mathcal{T} \overset{\text{LD}}{\rightsquigarrow} \mathcal{T}'$, when a compact neighbourhood $K \subset \mathbb{R}^d$ of 0 exists such that, whenever $(-x + \mathcal{T}) \sqcap K = (-y + \mathcal{T}) \sqcap K$ holds for $x, y \in \mathbb{R}^d$, one also has $(-x + \mathcal{T}') \sqcap \{0\} = (-y + \mathcal{T}') \sqcap \{0\}$.

Two patterns $\mathcal{T}_1, \mathcal{T}_2 \sqsubset \mathbb{R}^d$ are called *mutually locally derivable* (MLD) from each other when $\mathcal{T}_1 \overset{\text{LD}}{\rightsquigarrow} \mathcal{T}_2$ and $\mathcal{T}_2 \overset{\text{LD}}{\rightsquigarrow} \mathcal{T}_1$. Similarly, two LI classes are MLD when they are locally derivable from each other.

Corollary 4.2 (Corollary 5.2 in Ref. 8). *Two crystallographic, locally finite point sets $\Lambda, \Lambda' \subset \mathbb{R}^d$ are MLD if and only if they have the same lattice of periods.*

Mutual local derivability allows to move between different descriptions of a structure, for instance a tiling and a point set representative of the same MLD class. We will make use of this and choose the most convenient representative in some of the discussion below.

4.2. *Repetitivity and continuous hull*

A pattern $\mathcal{T} \sqsubset \mathbb{R}^d$ is called (translationally) *repetitive* when, for every compact $K \subset \mathbb{R}^d$, there is a compact $K' \subset \mathbb{R}^d$ such that, for every $x, y \in \mathbb{R}^d$, the relation $\mathcal{T} \sqcap (x + K) = (-t + \mathcal{T}) \sqcap (y + K)$ holds for some $t \in K'$. Choose $K = K_r := \overline{B_r(0)}$ and $K' = K_R$ such that $R \geq r$ is minimal. The function $R = R(r)$ is called the *repetitivity function*. A repetitive pattern $\mathcal{T} \sqsubset \mathbb{R}^d$ is called *linearly repetitive* when its repetitivity function satisfies $R(r) = \mathcal{O}(r)$ as $r \to \infty$.

Proposition 4.3 (Proposition 5.3 in Ref. 8). *Let $\mathcal{T} \sqsubset \mathbb{R}$ be a tiling that emerges via the geometric interpretation of a primitive substitution rule on a finite alphabet. Then, $\mathcal{T}$ is linearly repetitive.*

A pattern $\mathcal{T} \sqsubset \mathbb{R}^d$ has *finite local complexity*, or *FLC* for short, when, for every compact set $K \subset \mathbb{R}^d$, the set of K-clusters $\{(t + K) \sqcap \mathcal{T} \mid t \in \mathbb{R}^d\}$ consists of finitely many equivalence classes up to $\mathbb{R}^d$-translations.

If $\Lambda \subset \mathbb{R}^d$ is an FLC set, its geometric or *continuous hull* is $\mathbb{X}(\Lambda) = \overline{\{t + \Lambda \mid t \in \mathbb{R}^d\}}$, where the closure is taken in the local topology. If the $\mathbb{R}^d$-orbit of every element $\Lambda' \in \mathbb{X}(\Lambda)$ is dense, the hull $\mathbb{X}(\Lambda)$ is called *minimal*. The subset

$$\mathbb{X}_0(\Lambda) = \{\Lambda' \in \mathbb{X}(\Lambda) \mid 0 \in \Lambda'\},$$

is sometimes called the *discrete hull* or the *transversal*.

5. Inflation Tilings

We now consider tilings and their symmetries. It turns out that it is useful to generalise the symmetry concept, in the sense that you do not just look at a single tiling, but the entire LI class or hull of a tiling. We will then define local inflation deflation symmetry, which generalises the geometric inflation considered above, and provides a powerful tool to construct aperiodic tilings.

5.1. *Symmetry and inflation*

Previously, we considered rotational symmetries of lattice point sets. The following result shows that planar point sets showing a rotational symmetry can either have a single rotation centre or must be periodic.

Proposition 5.1 (Proposition 5.7 in Ref. 8). *Let $\Lambda \subset \mathbb{R}^2$ be a uniformly discrete point set with an exact n-fold rotational symmetry. If n is non-crystallographic, which means $n = 5$ or $n \geq 7$, there can only be one such rotation centre. When $n \in \{3, 4, 6\}$, the existence of more than one rotation centre is possible, and then implies lattice periodicity of Λ. When $n = 2$, the existence of another rotation centre means that Λ is at least rank-1 periodic.*

This shows that it is necessary to generalise the notion of symmetry to make sense for the non-crystallographic case. This is done by defining symmetry via local indistinguishability. Let R be a linear or affine transformation of $\mathbb{R}^d$. A pattern $\mathcal{T} \sqsubset \mathbb{R}^d$ is *symmetric* under the action of R when $R(\mathcal{T}) \overset{\text{LI}}{\sim} \mathcal{T}$. Moreover, the hull $\mathbb{X}(\mathcal{T})$ is *symmetric* under the action of R when $R(\mathbb{X}(\mathcal{T})) \subset \mathbb{X}(\mathcal{T})$. Note that this definition does not imply that there are tilings in the hull that individually possess this symmetry (in the usual sense), although there may be; we shall encounter an example of such a situation later on.

A discrete structure $\mathcal{T}$ in $\mathbb{R}^d$ is said to have a *local scaling property* with respect to the homothety $x \mapsto \lambda x$ for some $0 \neq \lambda \in \mathbb{R}$, if $\lambda \mathcal{T} \overset{\text{LD}}{\leadsto} \mathcal{T}$. A discrete structure $\mathcal{T}$ in $\mathbb{R}^d$ is said to have a *local inflation deflation symmetry* (LIDS) relative to the linear map L if $\mathcal{T} \overset{\text{MLD}}{\leadsto} L(\mathcal{T})$. When $L(x) = \lambda x$, the number λ is called the *inflation multiplier* of the LIDS.

Consider a finite set $\{T_1, T_2, \ldots, T_n\}$ of tiles, where each $T_i \subset \mathbb{R}^d$ is a compact set with non-empty interior and $\overline{T_i^\circ} = T_i$, so that we also have $0 < \mathrm{vol}(T_i) < \infty$. An *inflation rule* with inflation multiplier

$\lambda > 1$ (and an extension map $x \mapsto \lambda x$) consists of the mappings

$$\lambda T_i \longmapsto \bigcup_{j=1}^{n} T_j + A_{ji}$$

with finite sets $A_{ji} \subset \mathbb{R}^d$, subject to the mutual disjointness of the interiors of the sets on the right-hand side and to the (individual) volume consistency conditions $\mathrm{vol}(\lambda T_i) = \sum_{j=1}^{n} \mathrm{vol}(T_j)\,\mathrm{card}(A_{ji})$, both for each $1 \le i \le n$. The matrix M defined by $M_{k\ell} = \mathrm{card}(A_{k\ell})$ is called the *inflation matrix*. The consistency conditions mean that λ^d is the leading eigenvalue of M and that $\big(\mathrm{vol}(T_1), \ldots, \mathrm{vol}(T_n)\big)$ is a corresponding left eigenvector of M.

Example 5.2. The table tiling consists of two prototiles (rectangles) with inflation rule and matrix

$$\square \to \;\blacksquare\!\square\! \qquad \blacksquare \to \blacksquare \qquad \text{and} \quad M = \begin{pmatrix} 2 & 2 \\ 2 & 2 \end{pmatrix},$$

with leading eigenvalue $\lambda^2 = 4$ and eigenvector $(1,1)$. A patch of the corresponding tiling obtained by repeated inflation of a single tile is shown in Fig. 3.

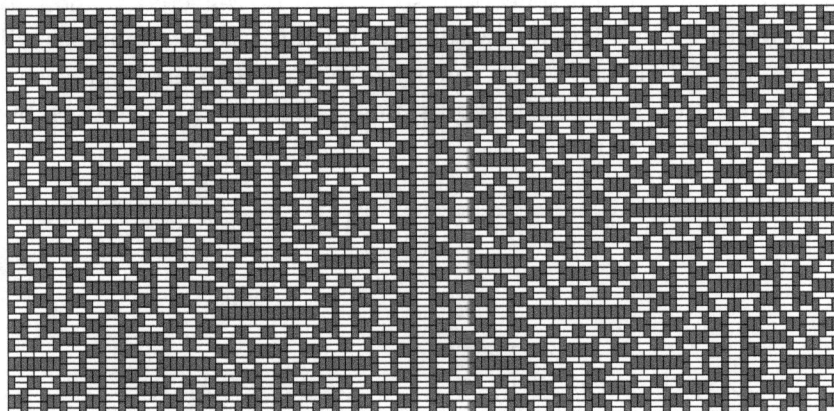

Fig. 3.   A patch of the table tiling.

The following result shows that inflation symmetry with an irrational inflation multiplier implies non-periodicity of a tiling.

Theorem 5.3 (Theorem 6.2 in Ref. 8). *Let $\mathcal{T} \sqsubset \mathbb{R}^d$ be an FLC pattern that satisfies the relation $\lambda \mathcal{T} \overset{LD}{\leadsto} \mathcal{T}$ for some $\lambda > 1$. If λ is irrational, the pattern $\mathcal{T}$ is non-periodic.*

We will now consider some important examples of tilings in detail.

5.2. *The Ammann–Beenker tiling*

We showed a tiling with eightfold symmetry at the beginning of this chapter, in Fig. 1. While this tiling is also based on an inflation rule, the best-known planar inflation tiling with octagonal symmetry is arguably the Ammann–Beenker tiling. While it is usually displayed as a tiling of squares and 45° rhombuses, it is convenient to use triangles and rhombuses as prototiles for an inflation approach. The inflation rule then employs two prototiles (up to Euclidean motion), a right isosceles triangle (which you may think of as half of a square) and a rhombus with a 45° angle. The inflation rule is shown in Fig. 4, the linear scaling factor is $\lambda = 1 + \sqrt{2}$. Note that the inflation of the triangle depends on the direction of the arrow decoration. As it is customary, it is implicitly understood that the inflation is compatible with rotation and reflection, so rotated or reflected versions of the prototiles are inflated accordingly. Within a tiling, triangles will always meet along their long side such that the two arrows match.

The inflation matrix (distinguishing triangles and rhombuses, but not their orientations) is

$$M = \begin{pmatrix} 3 & 4 \\ 2 & 3 \end{pmatrix}$$

Fig. 4.   The inflation rule of the Ammann–Beenker tiling.

with Perron–Frobenius eigenvalue $\lambda^2 = (1 + \sqrt{2})^2 = 3 + 2\sqrt{2}$. Its left eigenvector is $\frac{1}{2}(1, \sqrt{2})$, in accordance with the areas of the two prototiles, which are $\frac{1}{2}$ (triangle) and $\frac{1}{2}\sqrt{2}$ (rhombus), when choosing the edge length of the rhombus as 1. The right Perron–Frobenius eigenvector (in statistical normalisation, which means that the sum of its entries is 1) is given by $\left(2 - \sqrt{2}, \sqrt{2} - 1\right)^T$ and gives the frequencies of triangles and rhombuses in the infinite tiling. Multiplying by the areas of the tiles shows that triangles and rhombuses cover same area fraction in the Ammann–Beenker tiling.

In which sense does the inflation rule define an infinite tiling, and hence an LI class as well as a hull? Figure 5 shows successive inflation steps, staring from a (legal) square-shaped patch consisting of two triangles. After one inflation step, the same patch occurs in the centre of the inflated patch, but in a different orientation (rotated by 180°). However, after two steps, the same patch occurs in the centre of the inflated patch in the original orientation, and the same observation holds for subsequent patches in this sequence. This sequence defines two fixed point tilings under the square of the inflation rule — inflating twice does not change the original patch but extends it, and larger and larger parts of the tiling remain fixed, hence it converges to a

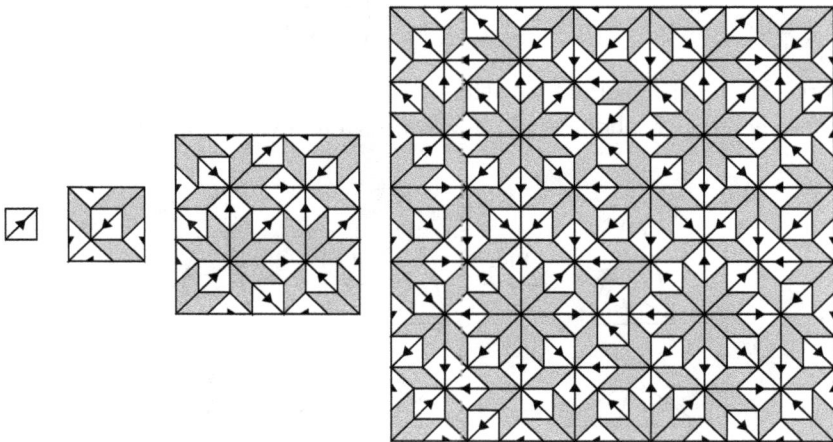

Fig. 5. Successive inflation steps for the Ammann–Beenker tiling.

fixed point. That the resulting tiling is non-periodic can already be seen from the fact that the ratio of squares to triangles is irrational, as follows from the analysis of the inflation matrix above.

The Ammann–Beenker tiling is thus a linearly repetitive tiling that is aperiodic and has finite local complexity. It possesses a local inflation deflation symmetry with inflation multiplier $1 + \sqrt{2}$.

5.3. *The Penrose and Tübingen triangle tilings*

The Penrose tilings[3] are arguably the most famous of all aperiodic tilings. They have tenfold symmetry, and there are many different versions, such as the pentagon and rhombic Penrose tiling. For our purposes, it is again more convenient to work with triangles, and we will describe a triangle version that is due to Robinson. It turns out that two very similar inflation rules produce the Penrose–Robinson tiling (PRT) as well as the Tübingen triangle tiling[10] (TTT), which are both built from the same triangular prototiles. Figure 6 shows both inflation rules.

The number of tiles in each inflated tiles is the same in both cases, the two rules only differ by the arrangement of triangles. Hence the corresponding inflation matrices are the same, and are given by

$$M = \begin{pmatrix} 2 & 1 \\ 1 & 1 \end{pmatrix} = \begin{pmatrix} 1 & 1 \\ 1 & 0 \end{pmatrix}^2.$$

Here M is the square of the inflation matrix of the Fibonacci sequence, hence the Perron–Frobenius eigenvalue of M is τ^2, and

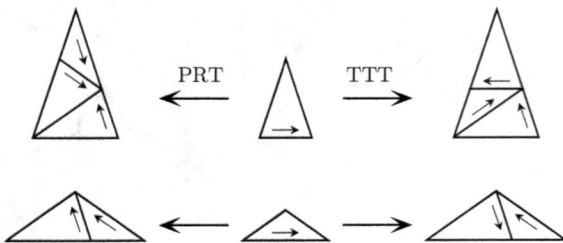

Fig. 6. The inflation rules of the Penrose–Robinson tiling (PRT) and the Tübingen triangle tiling (TTT).

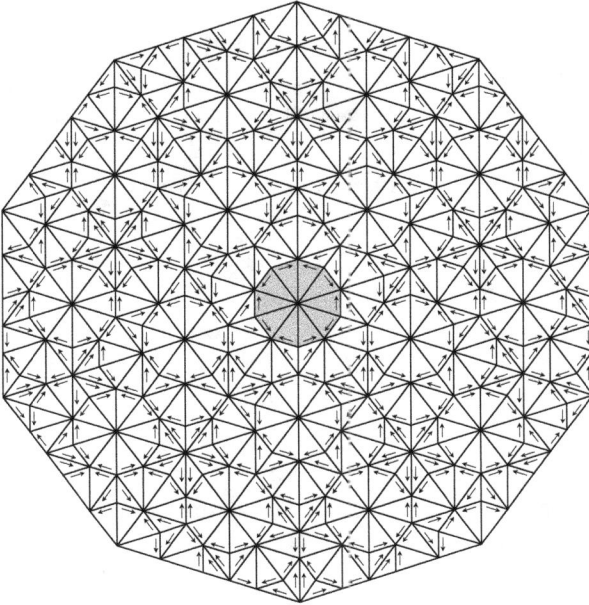

Fig. 7. A patch of the Penrose–Robinson tiling, obtained by four inflation steps from the central (shaded) decagon, which is a legal patch (you can see it occurring in the fourth inflation of a single triangle).

the left and right eigenvectors are the same as for the Fibonacci case. The area ratio of the prototiles is τ, and the frequency vector is $\nu = (\tau - 1, 2 - \tau)^T$. The larger triangles cover $\frac{1}{5}(2 + \tau) \approx 0.7236$ of the plane. A patch of the PRT is shown in Fig. 7. Due to the choice of seed, this particular tiling possesses perfect D_5 symmetry with respect to the centre point, where D_5 denotes the dihedral group of order 10. According to Proposition 5.1, there cannot be any other such centre in the entire (infinite) tiling. This also implies that in the hull there will be many tilings which have no such symmetry centres, because due to repetitivity you can create convergent sequences of tilings by shifting the symmetry centre out to infinity, and in the limit obtain a tiling without such a centre. Nevertheless, the tiling is symmetric in the sense defined above, namely that any rotation or reflection from D_5 maps the tiling to one that is locally indistinguishable.

The Penrose–Robinson tiling is a linearly repetitive tiling that is aperiodic, has finite local complexity and possesses a local inflation deflation symmetry with inflation multiplier τ. The hull possesses D_{10} symmetry, even though it does not contain any individual tiling with D_{10} symmetry; you can at most have D_5 symmetry of individual tilings. The various different versions of Penrose tilings (such as the rhombic Penrose tiling, kite and dart tiling or pentagon tiling) are all mutually locally derivable from the Penrose–Robinson tiling.

In contrast, there is no local rule to derive the Tübingen triangle tiling from the Penrose tiling, although there is a rule to do the converse. Hence these two tilings are *not* mutually locally derivable and define two distinct MLD classes. A patch of the Tübingen triangle tiling is shown in Fig. 8. Its hull also has D_{10} symmetry, but again does not contain any individually D_{10}-symmetric tilings. Indeed, there are no individually D_5-symmetric tilings in the TTT hull. Many more examples of inflation tilings can be found in the online Tilings Encyclopedia.[11]

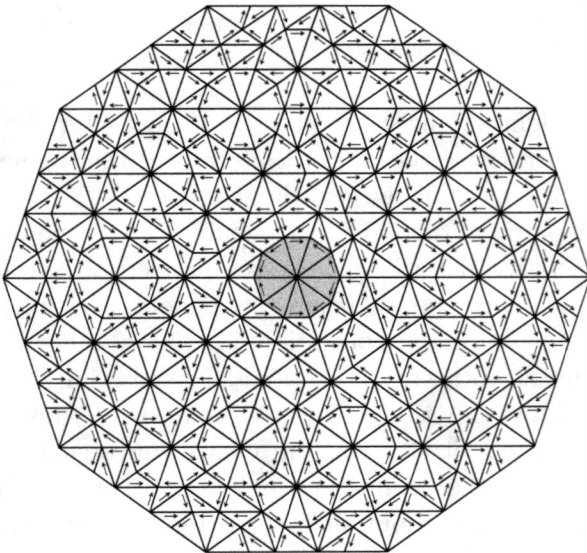

Fig. 8.   A patch of the Tübingen triangle tiling, obtained by four inflation steps from the central (shaded) decagon.

Exercise 5.4. The *chair tiling* consists of a single tile which occurs in four different orientations. The inflation rule for one orientation is

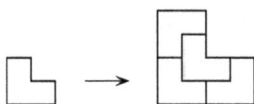

and the rules for the other three orientations are obtained by rotating both sides.

(a) Sketch the inflation rules for the remaining three orientations.
(b) Considering the four rotated versions of the prototile as separate tiles and write down the corresponding inflation matrix. If you start from a single tile in one orientation, how many tiles do you get after eight inflation steps in each of the four different orientations?
(c) Find the leading eigenvalue and the corresponding left and right eigenvector of the inflation matrix. Verify that all orientations occur equally often in an infinite fixed point tiling.
(d) Show that the symmetric patch

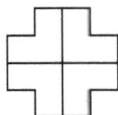

is legal for the chair inflation rule.
(e) Argue that inflation of the symmetric patch above leads to a fixed point tiling with individual fourfold symmetry.

6. Model Sets

In this section, we describe a different approach, which is based on a projection from a higher-dimensional periodic lattice. The resulting sets are called *cut and project sets* or *model sets*, and were first investigated by Meyer[12] in the context of harmonic analysis. It turns out that the examples discussed above allow for an alternative description via this approach (although, in general, inflation tilings are not model sets, and general model sets do not have a local inflation deflation symmetry). We shall explain the construction first for the example of the Fibonacci sequence and then concentrate on planar systems, making use of the cyclotomic fields and the Minkowski embedding discussed in Section 2. For further reading, we refer to Moody's review[13] and to Chapter 7 of Ref. 8.

6.1. *The Fibonacci model set*

Recall the Fibonacci inflation of Example 3.12. Starting from a legal patch consisting of two long intervals (a) with joint endpoint in the origin, application of the square of the inflation rule produces a fixed point tiling, corresponding to the bi-infinite fixed point sequence of Example 3.1. Call the set of all left endpoints of long intervals Λ_a and those of short intervals Λ_b, with $\Lambda = \Lambda_a \cup \Lambda_b$ being the set of all interval endpoints. The sets Λ_a, Λ_b and Λ are point sets in $\mathbb{R}$. In fact, because the two intervals have length τ (long interval) and 1 (short interval), all endpoints lie on integer linear combinations of these two values, hence are elements of $\mathbb{Z}[\tau] = \{m + n\tau \mid m, n \in \mathbb{Z}\}$. Recall the Minkowski embedding $\mathcal{L}$ of this set given in Eq. (2.3) and in Fig. 2. Since $\Lambda_a, \Lambda_b, \Lambda \subset \mathbb{Z}[\tau]$, the point sets lift to subsets of the lattice $\mathcal{L}$. Figure 9 shows what these subsets are. The corresponding lattice points lie within strips in the lattice $\mathcal{L}$, and *every* lattice point in these strips corresponds to an endpoint of an interval.

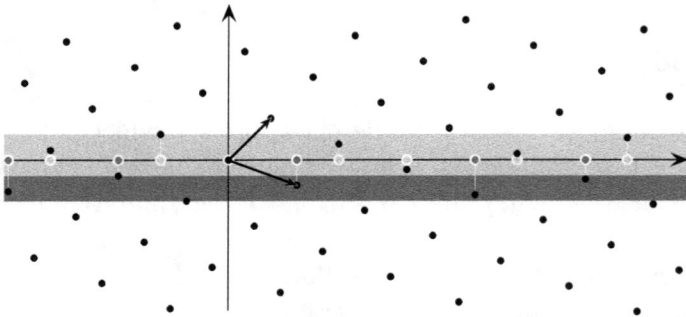

Fig. 9.   Lift of the Fibonacci point sets $\Lambda_{a,b} \subset \mathbb{Z}[\tau]$ to the Minkowski embedding $\mathcal{L}$.

Denoting the cross-sectional intervals of the strips corresponding to the lift of Λ_a and Λ_b by W_a and W_b, respectively, we can write the point sets as

$$\Lambda_{a,b} = \{x \in \mathbb{Z}[\tau] \mid x^\star \in W_{a,b}\}, \tag{6.1}$$

where the $\star$-map for an element $m + n\tau \in \mathbb{Z}[\tau]$ is nothing but algebraic conjugation,[a] so $(m + n\tau)^\star = m + n\tau' = m + n(1 - \tau) = m + n - n\tau$. So starting from x in the dense point set $\mathbb{Z}[\tau] \subset \mathbb{R}$, we lift it to $(x, x^\star)$ in the lattice $\mathcal{L} \subset \mathbb{R}^2$, and then select points within the strip by the condition that $x^\star \in W_{a,b}$. The intervals $W_{a,b}$ are called the *windows* of the cut and project sets Λ_a and Λ_b, and in our case are explicitly given by

$$W_a = (\tau - 2, \tau - 1] \quad \text{and} \quad W_b = (-1, \tau - 2]. \tag{6.2}$$

The set of all interval endpoints is $\Lambda = \Lambda_a \cup \Lambda_b = \{x \in L \mid x^\star \in W\}$ with the window $W = W_a \cup W_b = (-1, \tau - 1]$.

How do we find these windows, and show the connection? We are looking at a fixed point tiling under the square $\varrho^2 : a \mapsto aba, b \mapsto ab$ of the inflation rule, corresponding to a scaling factor τ^2, so Λ_a and Λ_b satisfy the set of set-valued equations

$$\begin{aligned}
\Lambda_a &= \tau^2 \Lambda_a \cup \tau^2 \Lambda_b \cup (\tau^2 \Lambda_a + \tau^2), \\
\Lambda_b &= (\tau^2 \Lambda_a + \tau) \cup (\tau^2 \Lambda_b + \tau).
\end{aligned} \tag{6.3}$$

Applying the $\star$-map on both sides and using $\Lambda_{a,b}^\star = W_{a,b}$ produces the following equations for the windows:

$$\begin{aligned}
\tau^2 W_a &= W_a \cup W_b \cup (W_a + 1), \\
\tau^2 W_b &= (W_a - \tau) \cup (W_b - \tau),
\end{aligned} \tag{6.4}$$

where all unions are disjoint. It is easy to check that the windows W_a and W_b of Eq. (6.2) satisfy these relations, as

$$\begin{aligned}
\tau^2(\tau - 2, \tau - 1] &= (-1, \tau] = (-1, \tau - 2) \cup (\tau - 2, \tau - 1] \cup (\tau - 1, \tau), \\
\tau^2(-1, \tau - 2] &= (-\tau - 1, -1] = (-\tau - 1, -2] \cup (-2, -1].
\end{aligned}$$

---

[a]We use the different notation $\star$ here because this is the conventional name for this map in this context.

Finally, one can check that the point sets obtained from inflation and the cut and project description have the same density and are repetitive.

6.2. *Euclidean and cyclotomic model sets*

The Fibonacci model set is an example of a general setup of a *model set* based on a *cut and project scheme*. This works in a general setting of locally compact Abelian groups, but here we limit ourselves to the case of *Euclidean model sets*. The general cut and project scheme is

$$
\begin{array}{ccccc}
\mathbb{R}^d & \xleftarrow{\;\;\pi\;\;} & \mathbb{R}^d \times \mathbb{R}^m & \xrightarrow{\;\;\pi_{\text{int}}\;\;} & \mathbb{R}^m \\
\cup & & \cup & & \cup \;\;\text{\scriptsize dense} \\
\pi(\mathcal{L}) & \xleftarrow{\;\;1\text{--}1\;\;} & \mathcal{L} & \xrightarrow{\hspace{2cm}} & \pi_{\text{int}}(\mathcal{L}) \\
\| & & & & \| \\
L & \xrightarrow{\hspace{3cm}\star\hspace{3cm}} & & & L^\star
\end{array}
\qquad (6.5)
$$

where $\mathcal{L}$ is a lattice in $\mathbb{R}^{d+m}$, with projections $L = \pi(\mathcal{L})$ to 'physical' space $\mathbb{R}^d$ and $L^\star = \pi_{\text{int}}(\mathcal{L})$ to 'internal' space $\mathbb{R}^m$. The $\star$-map relation is guaranteed by the requirement that the projection to physical space is one-to-one, so each point $x \in L$ has a unique well-defined partner $x^\star$ in internal space. A *Euclidean model set* then is any translate of the set $\Lambda = \{x \in L \mid x^\star \in W\}$, where the window W is a relatively compact subset of internal space $\mathbb{R}^m$ with non-empty interior. The model set Λ is a Meyer set; it is called a *regular model set* when the boundary of the window ∂W has zero Lebesgue measure, and a *generic (non-singular) model set* when $L^\star \cap \partial W = \varnothing$.

We are now discussing a particular class of Euclidean model sets which we call *cyclotomic model sets*. Here, we are looking at tiling in the plane $\mathbb{R}^2$, which we can identify with the complex plane $\mathbb{C}$; compare the discussion in Section 2.3 above. So, we are dealing with the case $d = 2$. Furthermore, the lattice $\mathcal{L}_n$ is the Minkowski embedding of the ring $\mathbb{Z}[\xi_n]$, where $\xi_n \in \mathbb{C}$ is a primitive nth root of unity. The dimension of the lattice is given by Euler's totient function $\phi(n)$ of Eq. (2.2), so the internal space has (real) dimension $\phi(n) - 2$. Hence the cut and project scheme has the form

$$
\begin{array}{ccccc}
\mathbb{R}^2 & \xleftarrow{\;\pi\;} & \mathbb{R}^2 \times \mathbb{R}^{\phi(n)-2} & \xrightarrow{\;\pi_{\text{int}}\;} & \mathbb{R}^{\phi(n)-2} \\
\cup & & \cup & & \cup \quad \text{dense} \\
\pi(\mathcal{L}_n) & \xleftarrow{\;1\text{-}1\;} & \mathcal{L}_n & \longrightarrow & \pi_{\text{int}}(\mathcal{L}_n) \\
\| & & & & \| \\
\mathbb{Z}[\xi_n] & & \xrightarrow{\quad\star\quad} & & \mathbb{Z}[\xi_n]^\star
\end{array}
\tag{6.6}
$$

with the $\star$-map defined as $x \mapsto (\sigma_2(x), \ldots, \sigma_{\frac{1}{2}\phi(n)}(x))$. Here, σ_i are the Galois automorphisms of $\mathbb{Q}(\xi_n)$, and the Minkowski embedding takes the form

$$
\mathcal{L}_n = \{(x, \sigma_2(x), \ldots, \sigma_{\frac{1}{2}\phi(n)}(x)) \mid x \in \mathbb{Z}[\xi_n]\} \subset \mathbb{C}^{\frac{1}{2}\phi(n)} \simeq \mathbb{R}^{\phi(n)}.
\tag{6.7}
$$

We shall now look at one (familiar) example of a cyclotomic model set in more detail.

6.3. *The Ammann–Beenker model set*

We choose the Minkowski embedding of $\mathbb{Z}[\xi_8]$, where we take the explicit choice $\xi_8 = e^{2\pi i/8}$ and the conjugation map defined by $\xi_8 \mapsto \xi_8^3$. This map acts on $\mathbb{Z}[\xi_8]$ by

$$
x = m_0 + m_1 \xi_8 + m_2 \xi_8^2 + m_3 \xi_8^3
$$
$$
\longmapsto x^\star = m_0 + m_3 \xi_8 - m_2 \xi_8^2 + m_1 \xi_8^3.
$$

The embedding leads to the lattice $\mathcal{L}_8 = \sqrt{2}\, R_8\, \mathbb{Z}^4$, with the rotation matrix

$$
R_8 = \frac{1}{2}
\begin{pmatrix}
\sqrt{2} & 1 & 0 & -1 \\
0 & 1 & \sqrt{2} & 1 \\
\sqrt{2} & -1 & 0 & 1 \\
0 & 1 & -\sqrt{2} & 1
\end{pmatrix}.
$$

Using a centred regular octagon O of unit edge length as the window, we obtain the vertex set of the Ammann–Beenker tiling as a model set. Explicitly, the vertex set is given by selecting any elements $x \in \mathbb{Z}[\xi_8]$ (which are of the form $x = m_0 + m_1 \xi_8 + m_2 \xi_8^2 + m_3 \xi_8^3$ with

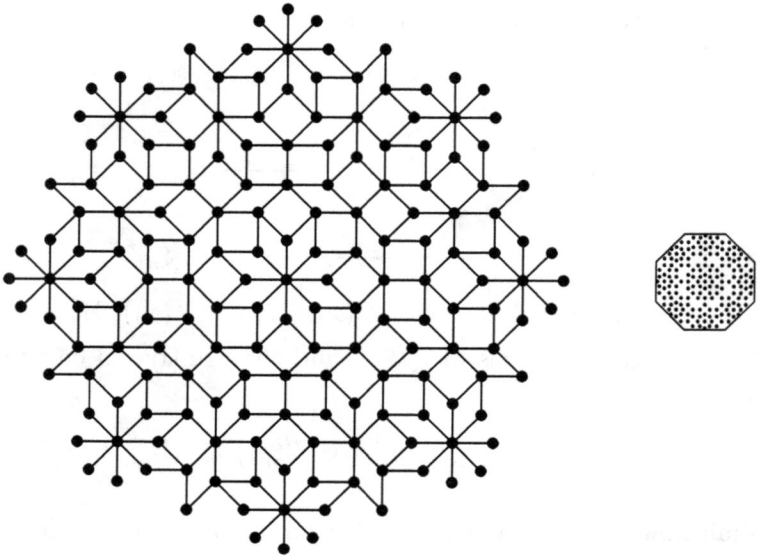

Fig. 10. Projection of the lattice $\mathcal{L}_8$ to physical space (left) and internal space (right). Note that the projection produces point sets; the lines on the left connect any two points of unit distance. Apart from the dissection of squares into two (oriented) triangles, this yields the same structure as the Ammann–Beenker inflation. In the limit of an infinite tiling, the points in internal space are uniformly distributed and fill the octagon densely.

$(m_0, m_1, m_2, m_3) \in \mathbb{Z}^4)$ whose image under the $\star$-map lies within the octagon O. A sketch of this construction is shown in Fig. 10.

The fact that the Ammann–Beenker tiling possesses a local inflation deflation symmetry can be seen from this description as well. The linear inflation multiplier (in physical space) is $\lambda = 1 + \sqrt{2}$. The corresponding action on the window is multiplication (scaling) by $\lambda^\star = -1/\lambda$. The rescaled octagon can be expressed as the intersection of eight translated copies of the original window, with translations that are elements of $\mathbb{Z}[\xi_8]$. Likewise, O can be written as a union of translated copies of the rescaled window $\lambda^\star O$; see Fig. 11. This implies that the model set and its λ-rescaled version are mutually locally derivable, so the model set possesses a local inflation deflation symmetry.

The vertex point set of the Tübingen triangle tiling can likewise be obtained, using the Minkowski embedding of $\mathbb{Z}[\xi_5]$, while the

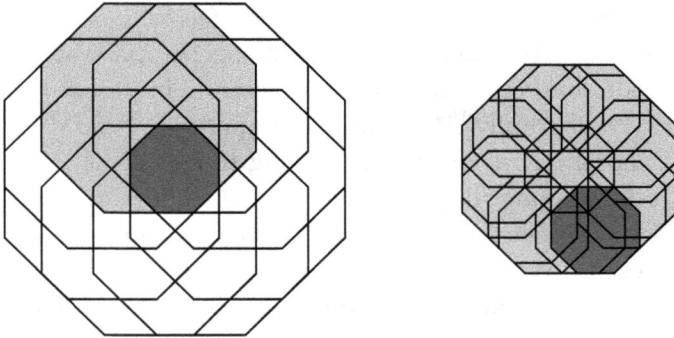

Fig. 11. Sketch of the reconstruction of the Ammann–Beenker tiling window. As shown in the left panel, the rescaled window (dark grey) can be obtained as the intersection of shifted copies of the original window (light grey). In the right panel, it is shown how the original window (light grey) is obtained as the union of shifted copies of the rescaled window (dark grey).

description of the Penrose tiling point set is a little more involved. For details and further examples we refer to Section 7.3 in Chapter 7 of Ref. 8.

Model sets are particularly nice, because (under some rather general assumptions) they are pure point diffractive and the associated dynamical system (with shift action on $\mathbb{R}^d$) has pure point dynamical spectrum. This is a consequence of the higher-dimensional periodicity. For a proof in a general setting we refer to Ref. 14; see also Chapter 9 in Ref. 8.

Exercise 6.1. Consider the silver mean $\lambda = 1 + \sqrt{2}$ and the silver mean model set $\Lambda = \{x \in \mathbb{Z}[\sqrt{2}] \mid x^\star \in W\}$, where the $\star$-map is algebraic conjugation $\sqrt{2} \mapsto -\sqrt{2}$ and the window $W = \left[-\frac{\sqrt{2}}{2}, \frac{\sqrt{2}}{2}\right]$. Find the elements of $x \in \Lambda$ with $|x| < 12$. Show that this list of points is consistent with the fixed point of the geometric version of the silver mean substitution rule $a \mapsto aba, b \mapsto a$ applied to the legal seed $a|a$.

7. Further Reading

The field of aperiodic order is still developing, and new insights are emerging. This chapter has only given a first glimpse at what has become a rich field with many difference facets. For instance, we did not discuss diffraction or dynamical spectra associated to tiling spaces; this has been a key area of research in recent years, and links to crystallography.[15–17] Spectra of aperiodic Schrödinger operators are another important field, linking to electronic properties of such materials. Again, important progress has been achieved, albeit mainly for one-dimensional systems. There has also been a considerable progress in the understanding of the topology of tiling spaces. These are just a few areas of current interest. Rather than pointing to specific publications covering these areas, we refer to the monograph[8] and to several collections of review articles,[18–21] as well as to the references contained therein.

Appendix A.   Solutions to the Exercises

Solution to Exercise 2.2.

(a) Clearly $\Lambda_a = 2\,\mathbb{Z}$ is uniformly discrete and relatively dense, so also Delone. All points are equivalent by translation so Λ_a is obviously FLC. As $\Lambda_a - \Lambda_a = \Lambda_a$ it is also Meyer.

(b) The set $\Lambda_b = \{n + 1/n \mid n \in \mathbb{Z}\backslash\{0\}\}$ is relatively dense as the maximum distance between points does not exceed 4. It is also uniformly discrete as points have minimum distance of $1/2$. The set Λ_b is not FLC because the distances between neighbouring points take on infinitely many values; and neither is it Meyer, because the difference set $\Lambda_b - \Lambda_b$ fails to be uniformly discrete. This can be seen, for instance, by looking at the distances between neighbouring points,

$$(n + 1) + \frac{1}{n + 1} - \left(n + \frac{1}{n}\right) = 1 + \frac{1}{n + 1} - \frac{1}{n} = 1 - \frac{1}{n(n + 1)},$$

so $\Lambda_b - \Lambda_b$ has an accumulation point at 1 (which itself is neither in Λ_b nor in $\Lambda_b - \Lambda_b$).

(c) The set $\Lambda_c = -\mathbb{N} \cup \{0\} \cup \sqrt{3}\,\mathbb{N}$ is clearly uniformly discrete and relatively dense, as point have minimum distance 1 and maximum distance $\sqrt{3}$, and hence also Delone. Moreover, it is clearly FLC as there are only finitely many local surrounding for any given finite radius. The difference set $\Lambda_c - \Lambda_c = \{m + n\sqrt{3} \mid$

$m, n \in \mathbb{Z}$ and $mn \geq 0\}$ fails to be uniformly discrete because $\sqrt{3}$ is irrational, so that

$$\inf\{|u - v| \mid u, v \in \Lambda_c - \Lambda_c, \, u \neq v\} = \inf\{|k - \ell\sqrt{3}| \mid k, \ell \in \mathbb{N}\}$$
$$= 0,$$

so Λ_c is not Meyer.

(d) Finally, consider $\Lambda_d = \mathbb{Z} \backslash S$, where S is an arbitrary subset of $2\mathbb{Z}$. Since all odd numbers are in Λ_d, it is clearly relatively dense. The minimum distance is still 1, so it is also uniformly discrete, and hence Delone. For any given radius, there are only finitely many ways in which points can be present or removed, so it is FLC. The difference set $\Lambda_d - \Lambda_d \subset \mathbb{Z}$, and actually equal to $\mathbb{Z}$ unless all odd points have been removed, in which case it is $2\mathbb{Z}$. So it is always uniformly discrete, and hence Λ_d is Meyer.

Solution to Exercise 2.6. Rotation by $\pi/5$ in $\mathbb{R}^2$ corresponds to multiplication by $\exp(\pi i/5)$ in $\mathbb{C}$, so the point set in $\mathbb{R}^2$ is invariant under rotation by $\pi/5$ if $\exp(\pi i/5)\,\mathbb{Z}[\xi] = \mathbb{Z}[\xi]$.

Observing that $-\exp(\pi i/5) = \exp(\pi i + \pi i/5) = \exp(6\pi i/5) = \xi^3$, and using that $\xi^5 = 1$ and $1 + \xi + \xi^2 + \xi^3 + \xi^4 = 0$, we find

$$-\xi^3\mathbb{Z}[\xi] = \{-a_0\xi^3 - a_1\xi^4 - a_2 - a_3\xi \mid a_0, a_1, a_2, a_3 \in \mathbb{Z}\}$$
$$= \{-a_0\xi^3 - a_1(-1 - \xi - \xi^2 - \xi^3)$$
$$\quad - a_2 - a_3\xi \mid a_0, a_1, a_2, a_3 \in \mathbb{Z}\}$$
$$= \{(a_1+a_3) + (a_1-a_3)\xi + a_1\xi^2$$
$$\quad + (a_1 - a_0)\xi^3 \mid a_0, a_1, a_2, a_3 \in \mathbb{Z}\}$$
$$= \mathbb{Z}[\xi],$$

where the final equality is true because the coefficients take all possible values in $\mathbb{Z}$.

Solution to Exercise 2.9. By definition, we have

$$\tau\,\mathbb{Z}[\tau] = \{m\tau + n\tau^2 \mid m, n \in \mathbb{Z}\} = \{m\tau + n\tau + n \mid m, n \in \mathbb{Z}\}$$
$$= \{n + (m+n)\tau \mid m, n \in \mathbb{Z}\} = \{n + \ell\tau \mid \ell, n \in \mathbb{Z}\} = \mathbb{Z}[\tau],$$

because $\tau^2 = \tau + 1$.

Solution to Exercise 2.10. The diagonal embedding is

$$\mathcal{L} = \{(x, x') \mid x \in \mathbb{Z}[\sqrt{2}]\} = \{(m + n\sqrt{2}, m - n\sqrt{2}) \mid m, n \in \mathbb{Z}\}$$
$$= \{m(1, 1) + n(\sqrt{2}, -\sqrt{2}) \mid m, n \in \mathbb{Z}\}$$

which is a planar lattice with basis vectors $v_1 = (1, 1)$ and $v_2 = (\sqrt{2}, -\sqrt{2})$. Since $|v_1| = \sqrt{2}$, $|v_2| = 2$ and $v_1 \cdot v_2 = 0$, $\mathcal{L}$ is a rotated rectangular lattice.

Solution to Exercise 3.2.

(a) We have $\varrho^2(a) = \varrho(abb) = abbaa$, so both letters occur in the $\varrho^2(a)$ and in $\varrho^2(b) = abb$, and hence ϱ is primitive.

(b) The substitution matrix for ϱ is

$$M = \begin{pmatrix} 1 & 1 \\ 2 & 0 \end{pmatrix}$$

which is not symmetric. The characteristic polynomial is $(1 - x)(-x) - 2 = x^2 - x - 2 = (x - 2)(x + 1)$, so the eigenvalues as $\lambda = 2$ and -1.

(c) The relation $w^{(n+1)} = w^{(n)} w^{(n-1)} w^{(n-1)}$ is proved by induction. Setting $w^{(0)} = a$, we have $w^{(1)} = \varrho(a) = abb$ and $w^{(2)} = \varrho(abb) = abbaa$. Clearly, these satisfy $w^{(2)} = w^{(1)} w^{(0)} w^{(0)}$. Assuming that $w^{(n+1)} = w^{(n)} w^{(n-1)} w^{(n-1)}$ holds, we find

$$w^{(n+2)} = \varrho(w^{(n+1)}) = \varrho(w^{(n)} w^{(n-1)} w^{(n-1)}) = w^{(n+1)} w^{(n)} w^{(n)},$$

which completes the proof.

(d) To compute the number of letters a and b in $w^{(8)}$ can be computed using the eighth power of the substitution matrix M. This can be calculated via

$$M^2 = \begin{pmatrix} 1 & 1 \\ 2 & 0 \end{pmatrix} \begin{pmatrix} 1 & 1 \\ 2 & 0 \end{pmatrix} = \begin{pmatrix} 3 & 1 \\ 2 & 2 \end{pmatrix},$$

$$M^4 = \begin{pmatrix} 3 & 1 \\ 2 & 2 \end{pmatrix} \begin{pmatrix} 3 & 1 \\ 2 & 2 \end{pmatrix} = \begin{pmatrix} 11 & 5 \\ 10 & 6 \end{pmatrix},$$

and hence

$$M^8 = \begin{pmatrix} 11 & 5 \\ 10 & 6 \end{pmatrix} \begin{pmatrix} 11 & 5 \\ 10 & 6 \end{pmatrix} = \begin{pmatrix} 171 & 85 \\ 170 & 86 \end{pmatrix}.$$

This gives

$$\begin{pmatrix} \mathrm{card}_a(w^{(8)}) \\ \mathrm{card}_a(w^{(8)}) \end{pmatrix} = M^8 \begin{pmatrix} 1 \\ 0 \end{pmatrix} = \begin{pmatrix} 171 & 85 \\ 170 & 86 \end{pmatrix} \begin{pmatrix} 1 \\ 0 \end{pmatrix} = \begin{pmatrix} 171 \\ 170 \end{pmatrix},$$

so there are 171 letters a and 170 letters b in $w^{(8)}$.

(e) The right eigenvector of M with eigenvalues $\lambda = 2$ is $(1,1)^t$, as

$$M \begin{pmatrix} 1 \\ 1 \end{pmatrix} = \begin{pmatrix} 1 & 1 \\ 2 & 0 \end{pmatrix} \begin{pmatrix} 1 \\ 1 \end{pmatrix} = \begin{pmatrix} 2 \\ 2 \end{pmatrix} = \lambda \begin{pmatrix} 1 \\ 1 \end{pmatrix}.$$

The letter frequencies in a fixed point word are encoded by the right eigenvector, so both letters are equally frequent. The corresponding ratio for $w^{(8)}$ was $\mathrm{card}_a(w^{(8)})/\mathrm{card}_b(w^{(8)}) = 171/170$.

Solution to Exercise 3.3.

(a) We have $\varrho^2(a) = abbaa$, which contains all combinations of two letters as subwords, so all four seeds $a|a$, $a|b$, $b|a$ and $b|b$ are legal.

(b) Applying ϱ to the four seeds gives

$a|a \mapsto abb|abb \mapsto abbaa|abbaa \mapsto cbbaaabbabb|abbaaabbabb \mapsto \ldots$

$a|b \mapsto abb|a \mapsto abbaa|abb \mapsto abbacabbabb|abbaa \mapsto \ldots$

$b|a \mapsto a|abb \mapsto abb|abbaa \mapsto abbaa|abbaaabbabb \mapsto \ldots$

$b|b \mapsto a|a \mapsto abb|abb \mapsto abbaa|abbaa \mapsto \ldots$

Clearly, there is no fixed point under ϱ, as the central two-letter words in all cases alternate between $a|a$ and $b|a$.

(c) The previous observation shows that there are two fixed points under ϱ^2, with cores $\ldots abbaa|abbaa \ldots$ and $\ldots bbabb|abbaaa \ldots$.

Solution to Exercise 3.13.

(a) The left eigenvector of the substitution matrix computed in Exercise 3.2 is $(2,1)$, because

$$(2,1) M = (2,1) \begin{pmatrix} 1 & 1 \\ 2 & 0 \end{pmatrix} = (4,2) = \lambda (2,1).$$

The appropriate interval length thus have length ratio 2:1, so in this case the longer interval has length 2. Indeed

$2 \times 2 = 4 = 2 + 1 + 1$ and $2 \times 1 = 2$, so this is consistent. The inflation rule is shown below.

(b) As both letters a and b are equally frequent, the distance between points in the set of left endpoints is either 1 and 2 with equal frequency, and so the mean distance is $3/2$. The density of the point set is the inverse (number of points per unit length), hence $2/3$.

Solution to Exercise 5.4.

(a) The three rotated inflation rules are

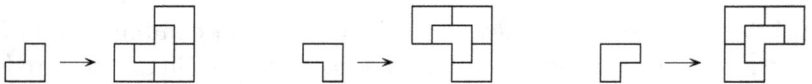

(b) As there are four different prototiles (up to translations), the inflation matrix is a 4×4 matrix. By inspection, it is given by

$$M = \begin{pmatrix} 2 & 1 & 0 & 1 \\ 1 & 2 & 1 & 0 \\ 0 & 1 & 2 & 1 \\ 1 & 0 & 1 & 2 \end{pmatrix}.$$

To calculate the number of tiles obtained in eight inflation steps, we need to compute the eighth power of this matrix,

$$M^2 = \begin{pmatrix} 2 & 1 & 0 & 1 \\ 1 & 2 & 1 & 0 \\ 0 & 1 & 2 & 1 \\ 1 & 0 & 1 & 2 \end{pmatrix} \begin{pmatrix} 2 & 1 & 0 & 1 \\ 1 & 2 & 1 & 0 \\ 0 & 1 & 2 & 1 \\ 1 & 0 & 1 & 2 \end{pmatrix} = 2 \begin{pmatrix} 3 & 2 & 1 & 2 \\ 2 & 4 & 2 & 1 \\ 1 & 2 & 3 & 2 \\ 2 & 1 & 2 & 3 \end{pmatrix},$$

$$M^4 = 4 \begin{pmatrix} 3 & 2 & 1 & 2 \\ 2 & 3 & 2 & 1 \\ 1 & 2 & 3 & 2 \\ 2 & 1 & 2 & 3 \end{pmatrix} \begin{pmatrix} 3 & 2 & 1 & 2 \\ 2 & 3 & 2 & 1 \\ 1 & 2 & 3 & 2 \\ 2 & 1 & 2 & 3 \end{pmatrix} = 8 \begin{pmatrix} 9 & 8 & 7 & 8 \\ 8 & 9 & 8 & 7 \\ 7 & 8 & 9 & 8 \\ 8 & 7 & 8 & 9 \end{pmatrix},$$

$$M^8 = 64 \begin{pmatrix} 9 & 8 & 7 & 8 \\ 8 & 9 & 8 & 7 \\ 7 & 8 & 9 & 8 \\ 8 & 7 & 8 & 9 \end{pmatrix} \begin{pmatrix} 9 & 8 & 7 & 8 \\ 8 & 9 & 8 & 7 \\ 7 & 8 & 9 & 8 \\ 8 & 7 & 8 & 9 \end{pmatrix}$$

$$= 128 \begin{pmatrix} 129 & 128 & 127 & 128 \\ 128 & 129 & 128 & 127 \\ 127 & 128 & 129 & 128 \\ 128 & 127 & 128 & 129 \end{pmatrix}$$

so there are $128 \times 129 = 16\,512$ tiles in the same orientation, $128^2 = 16\,384$ tiles rotated by $\pm\pi/2$, and $128 \times 127 = 16\,256$ tiles rotated by π.

(c) The matrix M is symmetric, so the left and right eigenvectors are transposes of each other. The leading eigenvalue is 4, with left eigenvector $(1,1,1,1)$ and right eigenvector $(1,1,1,1)^T$. As all entries are the same, all orientations occur with equal frequency in a fixed point tiling.

(d) The symmetric patch occurs in the threefold inflation of a single chair tile

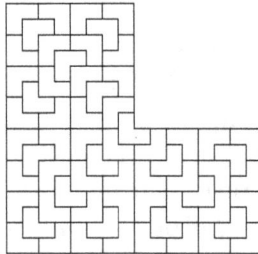

so it is legal for the inflation rule.

(e) The initial patch has fourfold rotation symmetry, and the inflation is compatible with rotations by multiples of $\pi/2$ by construction, so the image of the patch under any power of the chair inflation rule also has fourfold rotational symmetry. As the centre patch of the tiling stabilises under inflation (due to the fact that the inflated tile retains the original tile at its position), iteration of the chair inflation rule on the symmetric patch converges to a fixed point tiling with individual fourfold symmetry. In fact, as the patch is also reflection symmetric, and the reflection symmetry is also preserved by the inflation rule, the tiling has individual D_4 symmetry. A patch of the tiling is shown below.

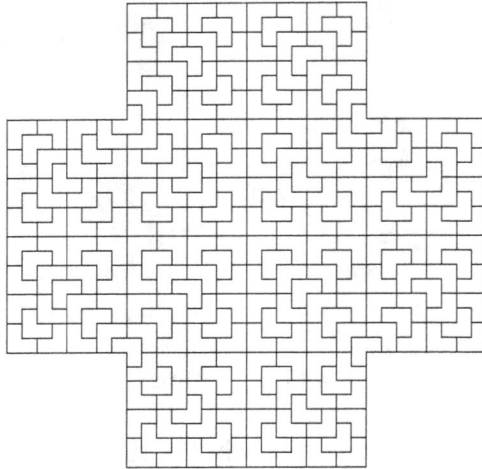

Solution to Exercise 6.1. The elements of $\mathbb{Z}[\sqrt{2}]$ have the form $x = m + n\sqrt{2}$ with $m, n \in \mathbb{Z}$. The image under the $\star$-map is $x^\star = m - n\sqrt{2}$, and so elements in Λ satisfy $-\sqrt{2}/2 \leq m - n\sqrt{2} \leq \sqrt{2}/2$. Clearly $m = n = 0$ satisfies this equation, and for every pair (m, n) (corresponding to a lattice point in the Minkowski embedding) that satisfies the inequality, the pair $(-m, -n)$ is also a solution. This implies that the point set Λ is symmetric about 0. Noting that 1 and $\sqrt{2}$ are both larger than $\sqrt{2}/2$, we see that m and n have to have the same sign (or $m = n = 0$), so we can limit ourselves to check positive values of m and n. Explicit calculation shows that the following pairs satisfy the relation $\{(0, 0), (1, 1), (2, 1), (3, 2), (4, 3), (5, 4), (6, 4), \ldots\}$ which gives (with $\lambda = 1 + \sqrt{2}$)

$$\{0, \pm\lambda, \pm(1 + \lambda), \pm(1 + 2\lambda), \pm(1 + 3\lambda), \pm(1 + 4\lambda), \pm(2 + 4\lambda)\} \subset \Lambda.$$

The natural interval lengths for the geometric realisation of the silver mean substitution are λ (for a) and 1 (for b). The fixed point sequence is ... *abaaaba|abaaaba* ..., in accordance with the distances between subsequent points in the list above.

References

1. D. Shechtman, I. Blech, D. Gratias and J. Cahn, Metallic phase with long-range orientational order and no translational symmetry, *Phys. Rev. Lett.* **53**, 1951–1953 (1984).
2. R. Berger, The undecidability of the domino problem, *Mem. Amer. Math. Soc.* **66**, 1–72 (1966).
3. R. Penrose, The rôle of aesthetics in pure and applied mathematical research, *Bull. Inst. Math. Appl.* **10**, 266–271 (1974).
4. Y. Watanabe, T. Soma and M. Ito, Nonperiodic tessellation with eightfold rotational symmetry, *Acta Crystallogr. A* **51**, 936–942 (1987).
5. M. Baake, U. Grimm and R. Moody, What is aperiodic order? preprint (2002), arXiv:math.HO/0203252.
6. M. Baake, D. Damanik and U. Grimm, Aperiodic order and spectral properties, preprint (2015), arXiv:1506.04978.
7. M. Baake, D. Damanik and U. Grimm, What is aperiodic order? *Notices Amer. Math. Soc.* **63**, 647–650 (2016).
8. M. Baake and U. Grimm, *Aperiodic Order. Vol. 1: A Mathematical Invitation.* Encyclopedia of Mathematics and Its Applications, Vol. 149, Cambridge University Press, Cambridge (2013).
9. J. Lagarias, Geometric models for quasicrystals I. Delone sets of finite type, *Discr. Comput. Geom.* **21**, 161–191 (1999).
10. M. Baake, P. Kramer, M. Schlottmann and D. Zeidler, Planar patterns with fivefold symmetry as sections of periodic structures in 4-space, *Int. J. Mod. Phys. B* **4**, 2217–2268 (1990).
11. D. Frettlöh, F. Gähler and E. Harriss, *Tilings Encyclopedia.* Available online at http://tilings.math.uni-bielefeld.de/.
12. Y. Meyer, *Algebraic Numbers and Harmonic Analysis.* North-Holland, Amsterdam (1972).
13. R. Moody. Model sets: a survey. In *From Quasicrystals to More Complex Systems*, eds. F. Axel, F. Dénoyer and J. Gazeau, pp. 145–166. EDP Sciences, Les Ulis (2000).
14. M. Schlottmann. Generalised model sets and dynamical systems. In *Directions in Mathematical Quasicrystals*, eds. M. Baake and R. Moody, pp. 143–159. American Mathematical Society, Providence, RI (2000).
15. M. Baake and U. Grimm, Kinematic diffraction from a mathematical viewpoint, *Z. Kristallogr.* **226**, 711–725 (2011), arXiv:1105.0095.
16. M. Baake and U. Grimm, Mathematical diffraction of aperiodic structures, *Chem. Soc. Rev.* **41**, 6821–6843 (2012), arXiv:1205.3633.
17. U. Grimm, Aperiodic crystals and beyond, *Acta Crystallogr. B* **71**, 258–274 (2015), arXiv:1506.05276.

18. R. Moody, ed., *The Mathematics of Long-Range Aperiodic Order.* NATO ASI Series C, Vol. 489, Kluwer Academic Publishers, Dordrecht (1997).
19. J. Patera, ed., *Quasicrystals and Discrete Geometry.* American Mathematical Society, Providence, RI (1998). Fields Institute Monographs, Vol. 10.
20. M. Baake and R. Moody, eds., *Directions in Mathematical Quasicrystals.* American Mathematical Society, Providence, RI (2000).
21. J. Kellendonk, D. Lenz and J. Savinien, eds., *Mathematics of Aperiodic Order.* Progress in Mathematics Vol. 309, Springer, Basel (2015).

Chapter 3

Complex Systems Dynamics

Hannah M. Fry

Centre for Advanced Spatial Analysis (CASA), University College
London, Gower Street, London, WC1E 6BT, UK
hannah.fry@ucl.ac.uk

Although a formal definition is yet to be agreed upon, complex systems
are often described as being composed of many interacting parts with
macro-level properties which do not follow trivially from the behaviour
of the individual components. Such systems can be found in biological,
economical, technological and social contexts and are an intriguing area
of study in the scientific and mathematical communities. In this chapter
we explore some of the major themes of complex systems and discuss
how to analyse such systems using tools originating in dynamical systems theory, statistical mechanics and network analysis. The chapter
is structured around a series of case studies that have been chosen for
their mathematical interest, the relevance of their results to real-world
problems and the wide range of techniques they employ in drawing connections between the behaviour of systems at various scales.

1.   What are Complex Systems?

1.1.   *Three categories of problems*

In 1948 Warren Weaver, an American scientist and mathematician,
wrote a paper in which he tried to summarise all systems of scientific
interest into three distinct categories.[5]

The first category, which he termed *Problems of Simplicity*,
describes systems with a small number of objects and well-defined
interactions. These are problems which respond well to analytical

treatment that generally require only a handful of variables to solve. An example might be the movement of a snooker ball on a table since it is fairly straightforward to write down a series of equations to predict the path of the ball and its interactions with the cushions. The name of the category here is not intended to suggest that the problems need to be simple, only the interactions. Difficult problems in rigid mechanics, structural engineering and rocket science would all count as problems of simplicity thanks to the small number of interacting objects.

Sticking with the example of the snooker ball, Weaver points out that while two or three balls on a snooker table might still fit into this category, the problem becomes much more difficult as you add more balls. Beyond three or four balls the problem quickly becomes unmanageable. Especially if, as a 1940s scientist, you only have traditional analytic tools at your disposal.

And yet, Weaver explains, if there were thousands, millions or billions of balls the problem would become much simpler as the methods of statistical mechanics would become applicable.

Of course, with millions of snooker balls in a system the individual path of one ball can no longer be traced. But other important questions can be answered: On average, how many balls will hit one side in a window of time? What is the average speed of the balls? How far does a ball move on average before it is hit by another? How many impacts per second does one ball make on average?

These methods apply when the balls, their positions and velocities are distributed in a random and disorganised way which leads to Weaver's term for this second category: *Disorganised Complexity.*

Examples of systems of Disorganised Complexity include gas and fluid dynamics and, extreme events aside, the call patterns in a mobile telephone company or the payouts of an insurance firm. At the micro-level — the level of fluid particles, individual phone calls or house fires — these systems are random and disorganised. But the disorder leads to a macro-level behaviour which is well behaved, easy to understand, quantify and predict.

At the point in history that Weaver wrote his paper (and arguably still) we had tools to deal with small simple problems and tools to deal with large, random and disorganised problems, but relatively little to tackle problems in the middle.

This middle region, where a large number of interactions order the system in an unpredictable way is where the field of complex systems lies. It is this middle region that Weaver named *Problems of Organised Complexity*.

1.2. *Complex Systems: a definition*

Although Weaver's problems of organised complexity, also known as *complex systems* have been of interest for several decades, a universally agreed upon technical definition of precisely what constitutes a Complex System does not yet exist. However, most researchers in the field would agree with Mark Newman's summary[7]:

"A complex system is a system composed of many interacting parts, which displays collective behaviour that does not follow trivially from the behaviours of the individuals."

There are a number of systems which fit this description. For instance, consider the coordinated, swarm-like behaviour of a flock of birds, known as *murmuration*. Understanding the connection between the micro-level and the macro-level behaviour of the individuals is a non-trivial problem. The many interacting individuals make the movement of the flock extremely difficult to predict. Questions like: Which way will the flock move next? Where will it be/what form will it take in T time steps from now? are difficult (if not impossible) to answer by following the behaviour of one individual fish/bird alone. Likewise, unlike in a box of gas particles, here each individual is capable of making decisions. This organisation at the micro-level means that traditional statistical mechanics methods might not be applicable.

The interacting parts of a complex system that Newman describes do not necessarily need to be autonomous individuals as they are in the case of murmuration. Take a bilingual dictionary, for example. It will contain a very large set of one to many mappings. If all that was required was word-for-word translation, the problem would be relatively simple. However, the structure of language: sentences, grammar and so on organises the components, and generates a wide number of rules and interactions between the words. On a small scale, with a very limited vocabulary, translating between simple sentences accurately is a manageable task, but on the scale of an entire language the problem becomes extremely complex.

Generally speaking, complex systems are seen to exhibit some or all of the following: *Feedback*; *Emergence*; *Lack of Central Control* and *Nonlinearity*. However, none of these are sufficient to define a complex system and arguably, many are not necessary either. The debate about a quantitative classification is even murkier. See Ref. 6 for an excellent overview of the main points of discussion.

1.3. *How to tackle complex systems*

What researchers can agree on, however, are the class of methods available to study these problems. In an effort to forge a relationship between the micro-scale behaviour and the macro-level properties of the system much of the research in this area involves approaching Weaver's problems of organised complexity via approximating the system as either one of his other two categories.

This could be by taking the most important qualitative aspects and describing them analytically, by using a few variables and simple interactions and then translating into a solvable framework. This is generally the motivation behind the class of rule-based simulation(s) as we shall see in Section 2. It also underpins the objectives of complex network theory, a topic we explore in Section 3. Equally, this is the aim when researchers use dynamical systems theory to study the key characteristics of complex systems. Hence many of the methods outlined in other chapters of this book can also be relevant to problems in this area.

Alternatively a complex system can, on occasion, be approximated by a statistical process, at which point the tools of statistical mechanics and information theory come into play. An example of this kind, revolving around entropy maximisation methods, is included in Section 4.

Complex systems as a research area has only come into the spotlight very recently, and is highly interdisciplinary in nature. As a result, to quote Doyne Farmer: *"Complex systems theory is not a monolithic body of knowledge, but more a series of short stories than a novel"*. This chapter aims to reflect this by giving you a flavour of the broad research approaches rather than a comprehensive overview of the subject.

2. Rule-Based Simulation

The aim here is to create a set of rules (usually for a computer) which replicate the behaviour observed in the system. Once the model is built, you then sit back and observe the macro-level emergent behaviour of your system. The Game of Life is one particularly famous example of this type of model, although not motivated by a real-world problem.

If required, these rule-based models can incorporate a great deal of detail at the level of the individual. The class of rule-based simulation in which independently acting (autonomous) agents interact with one another and their environment are known as *agent-based* or *individual-based* models. This ability to include detail at the micro-level is one reason agent-based models have become popular in an interdisciplinary context, particularly when generating a realistic simulation of the individual behaviour and a qualitative understanding of the emergent properties is the main motivation.

From a mathematical perspective however, the most interesting agent-based models are those which go beyond qualitative descriptions and forge an analytical relationship between the micro- and macro-levels of a system. One such example of an agent-based system is that of rabbit and fox populations.

This model is motivated by the coupled Lotka–Volterra equations. These were studied in Chapter 1, but repeated here for convenience:

$$\frac{dR}{dt} = aR - bRF, \tag{2.1}$$

$$\frac{dF}{dt} = cRF - dF. \tag{2.2}$$

The idea behind the agent-based version is to replicate the general dynamics in (2.2) by allowing rabbits and foxes to roam around a two-dimensional space and act operating according to a simple set of rules.

In the simplest case, the model is based on a rectangular lattice with periodic boundary conditions and cell separation n. Each cell, C, on the lattice has three possible states: E when the cell is empty; R when it contains one rabbit; F when it contains one fox. In this version of the model, the following parameters govern the system:

α: The rabbit reproduction rate (if the rabbit is not over crowded).
β: The fox reproduction rate (if the fox has just eaten).
γ: The fox death rate.

The algorithm for the agent-based model is as follows:

At each point in time, a cell C is randomly selected. Then, one of C's eight immediate neighbours, labelled C', is also selected at random. The six possible states of C and C' determine what happens next.

If:

$C = R$, $C' = E$: The rabbit reproduces into C' with probability α.
$C = R$, $C' = F$: F eats R & reproduces into empty cell with probability β.
$C = F$, $C' = R$: As above.
$C = F$, $C' = E$ or $C' = F$: The fox dies with probability γ.
$C = E$, $C' = R$ or $C' = F$: The individual in C' moves to C.
$C = C' = R$ or $C = C' = E$: No change.

An example code implementing this algorithm may be found at www.hannahfry.co.uk/rabbitfoxes.

For suitably chosen values of the parameters, as in Fig. 1, this agent-based simulation does indeed demonstrate the cyclic behaviour seen in the Lotka–Volterra system.

But we can go further. With this formulation, it is also possible to write down the probabilities of rabbit and fox births. If $r(t)$ is the proportion of cells with rabbits at time t, $f(t)$ is the proportion of

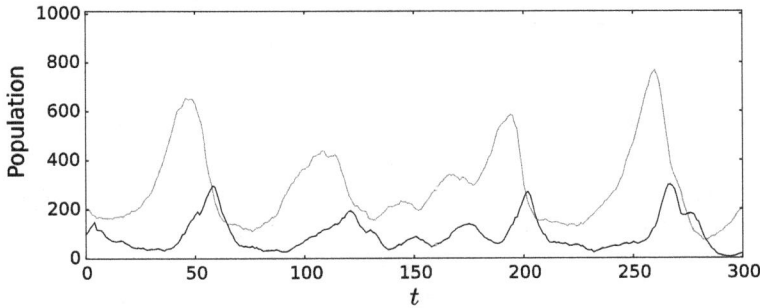

Fig. 1. Agent-based simulation of rabbit and fox populations demonstrating the cyclic behaviour in rabbit and fox populations seen in the Lotka–Volterra model.

cells containing foxes at time t then the proportion of empty cells will be $(1 - r - f)$. The probabilities per unit space in a unit time will be[10]:

Event	Probability
Rabbit born	$\delta t\, \alpha r(1 - r - f)/(n^2)$
Rabbit eaten	$\delta t\, rf/(n^2)$
Fox born	$\delta t\, \beta rf/(n^2)$
Fox eaten	$\delta t\, \gamma f(1 - r)/(n^2)$

These probabilities lead directly to the difference equations for the rabbit and fox populations:

$$r(t + 1) = r(t) + \frac{\delta t}{n^2} \left(\alpha r(1 - r - f) - rf \right), \tag{2.3}$$

$$f(t + 1) = f(t) + \frac{\delta t}{n^2} \left(\beta rf - \gamma f(1 - r) \right). \tag{2.4}$$

Taking the limit as $\delta t, n^2 \to \infty$ such that $\delta t/n^2 = O(1)$, we end up with a coupled set of differential equations:

$$\frac{1}{u} \frac{\partial r}{\partial t} = \alpha - \alpha r + (1 + \alpha)f, \tag{2.5}$$

$$\frac{1}{f} \frac{\partial f}{\partial t} = r(\beta + \gamma) - \gamma. \tag{2.6}$$

These are almost equivalent to the Lotka–Volterra equations in (2.2) when $\alpha = a$, $1 + \alpha = b$, $\beta + \gamma = c$, $\gamma = d$, except here there is an additional $-\alpha r^2$ term in the equation for rabbits. This is owing to the population being limited by the number of free cells (a rabbit can only reproduce if its neighbouring cell is empty).

This example demonstrates that it is possible to create an analytical understanding of how the inputs to a simulation at the micro-level will affect the dynamics at the macro-level, but these examples are few and far between with agent-based models. Finding a universal understanding of how to translate between different scales within simulation is an active area of research. For that reason we leave this section brief and invite interested readers to refer to Refs. 9 and 10.

3.   Networks

If complex systems are fundamentally comprised of interacting individuals, any study of the interactions and how they unfold across the system must first require a good understanding of the map of interactions. This underlying network and its topology of connections will play a key role in the feedback and emergence of the system. Exploring how is, at least in part, the motivation behind the recent surge in network science.

There are several obvious examples of networks within complex systems. (1) The street network and crime across a city. (2) The world wide web and its influence on communication and information flow. (3) Telephone networks, transport networks, social networks. But there are also some that are a little less obvious. The synapses within the brain; the influence between fireflies flashing in sync; the connections between words within a sentence. All play a major part on the emergent properties of their respective systems.

Although graph theory is certainly not a new subject area, traditionally it focused on regular or random networks which, as we shall see, have quite different characteristics from those found in many complex systems. Nonetheless, the notation and definitions of graph

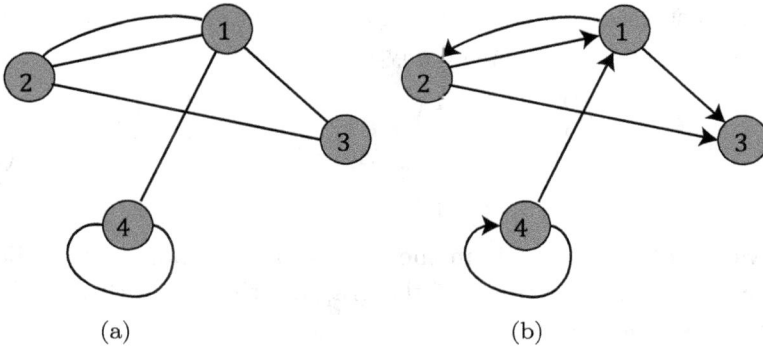

Fig. 2. Example networks: (a) G_1 — an undirected network; (b) G_2 — an directed network.

theory are essential to describing and analysing complex networks, and so this is where we begin.

3.1. *Elementary graph theory*

A *graph*, or *network*, is an object comprised of a series of *nodes* which are connected by *edges*. (Nodes and edges are also known as *vertices* and *links* respectively and we will use these terms interchangeably.) Networks can be *undirected*, as in Fig. 2(a). Or, if the edges indicate the direction of a path between two nodes, the network is unsurprisingly known as *directed*. An example of a directed graph is given in Fig. 2(b).

The *adjacency matrix* of a finite network is a matrix which indicates the topology of the network. Unless there are any self-loops in the network, it will have zeros in the diagonal. An undirected network will have a symmetric adjacency matrix. Convention dictates which way you define the adjacency matrix for directed networks. Here we will use the following:

$$a_{ij} = \begin{cases} 1 & \text{if there is an edge from node } j \text{ to node } i, \\ 0 & \text{otherwise.} \end{cases} \tag{3.1}$$

This configuration is fairly standard throughout the network literature. (Although, confusingly, the exact opposite definition is also used by some.)

For illustration, here are the adjacency matrices for the networks in Figs. 2(a) and 2(b) respectively:

$$A_1 = \begin{pmatrix} 0 & 2 & 1 & 1 \\ 2 & 0 & 1 & 0 \\ 1 & 1 & 0 & 0 \\ 1 & 0 & 0 & 1 \end{pmatrix}, \quad A_2 = \begin{pmatrix} 0 & 1 & 0 & 1 \\ 1 & 0 & 0 & 0 \\ 1 & 1 & 0 & 0 \\ 0 & 0 & 0 & 1 \end{pmatrix}. \tag{3.2}$$

The values of the elements in the matrix can be adopted to indicate a weighting along the edges of the graph, although in (3.2) we have taken all weights to be unity.

The *degree* of a node is the number of edges per node (these will be oriented in the case of a directed graph). The degree is usually denoted by k and can be easily determined from the adjacency matrix.

3.1.1. *Degree in undirected graphs*

In an undirected graph, the degree k_i is simply the sum of the ith row of the adjacency matrix:

$$k_i = \sum_{j=1}^{n} A_{ij}. \tag{3.3}$$

The total degree of graph G is twice the total number of edges (since in an undirected graph each edge will count twice):

$$2m = \sum_{i=1}^{n} k_i = \sum_{ij} A_{ij}. \tag{3.4}$$

Together these give a neat expression for the mean degree in a graph:

$$c = \frac{1}{n} \sum_{i=1}^{n} k_i = \frac{2m}{n}. \tag{3.5}$$

The maximum possible number of edges in a graph is $n(n-1)/2$, which means we can also define the *density* of the graph as the ratio of edges to the maximum number possible:

$$\rho = \frac{2m}{n(n-1)}. \tag{3.6}$$

When $\rho = 1$ the graph is fully connected.

3.1.2. *Degree in directed graphs*

In directed networks, the degree of a node is split into *in-degree* and *out-degree*. In-degree for node i is the sum of the rows of the adjacency matrix:

$$k_i^{\text{in}} = \sum_{j=1}^{n} A_{ij}. \tag{3.7}$$

Out-degree for node j is the sum of the columns of the adjacency matrix:

$$k_j^{\text{out}} = \sum_{i=1}^{n} A_{ij}. \tag{3.8}$$

The sums of the in-degree and out-degree over all nodes of the network both give the number of edges in the graph:

$$m = \sum_{i}^{n} k_i^{\text{in}} = \sum_{j}^{n} k_j^{\text{out}} = \sum_{ij}^{n} A_{ij}, \tag{3.9}$$

which together give an expression for the mean degree (in and out) and the density of the graph:

$$c = \frac{m}{n}, \quad \rho = \frac{m}{n(n-1)}. \tag{3.10}$$

3.2. *Network measures*

Beyond the degree and density of nodes in a network, there are several other useful measures which allow us to classify the different characteristics of graphs.

3.2.1. *Clustering Coefficient*

First up, the *clustering coefficient*. This is a way to measure how clique-y the network is by asking how many of your neighbours are also neighbours.

Consider node v with degree k_v in an undirected network G. Label the number of edges that exist between the neighbours of v as N_v.

The clustering coefficient $C_c(v)$ is simply a fraction of all possible edges between the neighbours of v that actually appear in the network G:

$$C_c(v) = \frac{2N_v}{k_v(k_v - 1)}. \tag{3.11}$$

So, in the example of the undirected network G_1 in Fig. 2(a), $C_c(1) = 1/3$ since only one of the possible three connections exists between its neighbours.

The clustering coefficient of a network is just the average clustering coefficient of all the nodes in the graph, calculated node by node. A fully connected graph will have $C_c(G) = 1$, while a network in the shape of a star with v in the centre, where no neighbours of v have connections between themselves, will have $C_c(G) = 0$.

3.2.2.   Centrality

Networks also have a series of measures to capture the importance of a particular node and its centrality in the graph. A node is considered *central* if:

- It has a high degree.
- It is easily accessible from all nodes.
- It lies on the shortest path between nodes.

These three definitions lead to the three main centrality measures:

Degree Centrality

A simple one to begin with. The *degree centrality* of a node v in an undirected network is simply the number of edges which connect to v, divided by the maximum number of nodes in the network v could be linked to, $n - 1$:

$$C_D(v) = \frac{\deg(v)}{n - 1}. \tag{3.12}$$

Closeness Centrality

The *network distance* d_{vw} is the number of hops in the path between two nodes v and w in the graph. The *shortest path*, σ_{vw}, is the

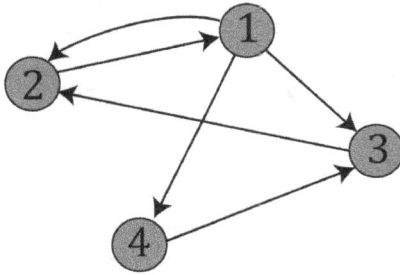

Fig. 3. G_3.

path between v and w such that no shorter path exists. (Again, pretty obvious.) For illustration, here is d_{vw} for the network G_3 in Fig. 3:

$$\sigma_{vw} = \begin{pmatrix} 0 & 1 & 1 & 1 \\ 1 & 0 & 2 & 2 \\ 2 & 1 & 0 & 3 \\ 3 & 2 & 1 & 0 \end{pmatrix}. \tag{3.13}$$

The average shortest path in a network is given by:

$$l_v = \frac{1}{n} \sum_{w=1}^{n} d_{vw}. \tag{3.14}$$

And the *closeness centrality* of node v, is then the reciprocal of the sum of the distances to all other $n-1$ nodes:

$$C(v) = \frac{n-1}{\sum_{w=1}^{n} d_{vw}}. \tag{3.15}$$

Betweenness Centrality

The *betweenness centrality* of a node v is the number of shortest paths from all vertices to all others which pass through the node in question:

$$g(v) = \sum_{i \neq v \neq j} \frac{\sigma_{ij}(v)}{\sigma_{ij}}, \tag{3.16}$$

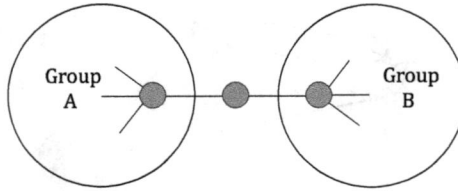

Fig. 4. A sketch of a network with high betweenness centrality and low degree centrality.

where σ_{ij} is the shortest path between every pair of nodes i and j in the network, and $\sigma_{ij}(v)$ are the shortest paths which pass through v.

The betweenness centrality has to be calculated by keeping a tally of all of the shortest paths between every node in the network. This can be a computationally expensive task, but the betweenness captures something quite different to the other two measures, as illustrated in the example sketched in Fig. 4. Betweenness has also been shown to be an important measure in real-world networks. For instance, in street networks betweenness correlates well with the footfall of a street segment and hence can be used to give an indication of how people might move around a city as pedestrians, without having to collect data.

3.3.   *Degree distribution*

A final, and important measure is the *degree distribution*. As you would expect, the degree distribution gives you insight into the spread of connections in the network and macro-level view of the topology.

The degree distribution was also the first measure to hint that the structure of networks in complex systems was fundamentally different to those that had been studied in traditional graph theory.

In 1959, Paul Erdos suggested that *random graphs*, that is, those with a Poisson degree distribution, can be used to describe the networks seen in communication and the life sciences. Random graphs do demonstrate the short path lengths that are seen in small world networks (think six degrees of separation) but they are also totally democratic. The variance in degree is very small and most nodes have approximately the same number of links.

In 1999 Albert Barabasi and colleagues tried to map the network of the world wide web. They expected, as Erdos had predicted, to find a random graph. After all, people have different tastes and interests, and so surely websites are picked virtually at random.

Instead the team found that 80% of all sites had less than four links, but a small number, 0.01% had more than 1000. To draw an analogy with a human system with a Gaussian distribution like height, finding these extremely well-connected nodes in the world wide web would be the equivalent of finding people who were 100ft tall.

This network of pages did not have an equal probability of finding a link between any two nodes. Instead the world wide web was found to have a power law for its degree distribution:

$$p_k = Ck^{-\alpha}, \tag{3.17}$$

for some constants C and α. Since power laws have the same functional form at all scales, this network became known as a *scale-free network*. Figure 5 shows a scale-free network and its degree distribution against that of a 2D lattice and random graph for comparison.

3.4. *Scale-free networks*

Since the 1999 survey of the internet by Barabasi and colleagues, scale-free networks have been shown to exist in a variety of real-world settings including social and collaboration networks, communication networks and semantic networks. They are characterised by their power law degree distribution, which we can use to explore some of their properties.

The degree distribution in (3.17) leads to a cumulative degree distribution:

$$P_k = \sum_{\hat{k}=k}^{\infty} p_{\hat{k}} = \sum_{\hat{k}=k}^{\infty} C\hat{k}^{-\alpha}. \tag{3.18}$$

H. M. Fry

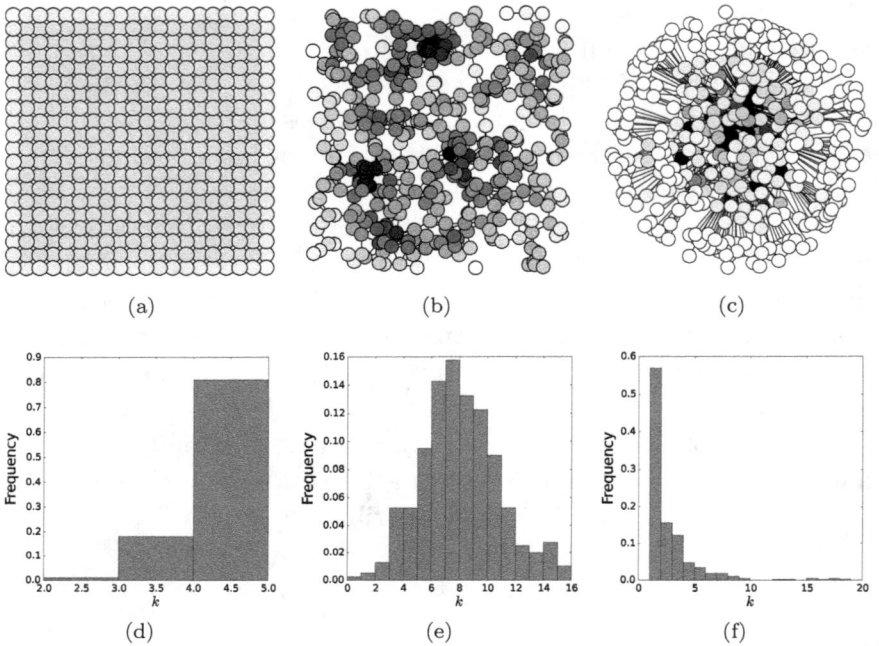

Fig. 5. Comparison between various networks and their degree distributions. Nodes in the top row are coloured by degree. (a) 2D lattice; (b) random graph; (c) scale free network; (d) degree distribution for the 2D lattice; (e) degree distribution for the random graph; and (f) degree distribution for the scale free network.

Approximating (3.18) as an integral with $d\hat{k} = 1$,

$$P_k \simeq \int_k^\infty C\hat{k}^{-\alpha} d\hat{k} \tag{3.19}$$

$$= \left[\frac{C\hat{k}^{-\alpha+1}}{-\alpha+1}\right]_k^\infty = \frac{C^{-(\alpha-1)}}{\alpha-1}. \tag{3.20}$$

Taking the log of the above:

$$\ln P_k = -(\alpha-1)\ln k + \ln C - \ln(\alpha-1). \tag{3.21}$$

Hence you can expect a scale-free network to form a straight line on a log-log scale with gradient $-(\alpha-1)$.

Since $P_{k\min} = 1$, that is, the probability of finding a node of the same size or bigger than the smallest within the network must be 1, we may write C in terms of $k_{\min}$:

$$C = (\alpha - 1)k_{\min}^{(\alpha-1)}. \tag{3.22}$$

Hence the cumulative the distribution is simply:

$$P_k = \left(\frac{k}{k_{\min}}\right)^{-(\alpha-1)}. \tag{3.23}$$

Notice the issues here when $k_{\min} = 0$. These also appear in the probability form:

$$p_k = Ck^{-\alpha} = \frac{C}{k^\alpha}. \tag{3.24}$$

If C is finite (which we want it to be for sensible values of $P_{k>0}$) then p_0 will be infinite. This cannot be since p_k is a probability. Thus, these power law descriptions only work for $k \geq 1$, and unconnected nodes are not considered part of the network.

Given C, p_k could also be written as

$$p_k = \frac{\alpha - 1}{k_{\min}}\left(\frac{k}{k_{\min}}\right)^{-\alpha}. \tag{3.25}$$

And since we require the sum of all $p_k = 1$:

$$\sum_{k=1}^{\infty} p_k = 1 = C\sum_{k=1}^{\infty} \frac{1}{k^\alpha}. \tag{3.26}$$

The summation on the right-hand side here is the Riemann zeta function $\zeta(\alpha) = 1/C$, which is convergent for $\alpha > 1$.

The first and second moments of this distribution of p_k are given by:

$$\langle k \rangle = \sum_{k=1}^{\infty} kp_k, \quad \langle k^2 \rangle = \sum_{k=1}^{\infty} k^2 p_k \tag{3.27}$$

and the variance:

$$\text{Var}(k) = E[(k - \langle k \rangle)^2] \tag{3.28}$$

$$= E\left[k^2 - 2k\langle k \rangle + \langle k \rangle^2\right] \tag{3.29}$$

$$= E(k^2) - 2\langle k \rangle E(k) + \langle k \rangle^2 \tag{3.30}$$
$$\sigma^2 = \langle k^2 \rangle - \langle k \rangle^2, \tag{3.31}$$

is the second moment minus the first moment squared. Here σ is the standard deviation of the distribution.

Most real-world scale-free networks seem to settle on α between $\alpha = 2$ and $\alpha = 3$, and as a result have a finite first moment, but a variance that tends to infinity. This characteristic leads to some interesting properties, as we shall see in Section 3, but first it is worth briefly pausing to discuss how these scale-free networks are created.

3.4.1. *Preferential attachment*

Put simply, scale-free networks can be generated whenever there is an underlying 'rich get richer' addition of new edges to a network. The precise process, which has become known as *preferential attachment* was first considered by Yule back in 1925, and then formalised to apply to a network by Price in 1976. However, the Barabasi–Albert formulation is the one most cited in the literature and the version we outline here.

The generation of a scale-free network begins with a graph of m_0 nodes. At each time step, a new node is added with $m \leq m_0$ edges. These edges are preferentially attached to existing nodes with a high degree. Specifically, the probability of a new edge attaching to existing node i is:

$$p_i = \frac{k_i}{\sum_j k_j}, \tag{3.32}$$

for degree k. After t time steps, the network will have $n = m_0 + t$ nodes and mt edges, and the resulting degree distribution can be found as follows.

Assume that k is continuous and k_i is the degree of node i. Then:

$$k_i(t + \delta t) = k(t) + \delta t A \frac{k_i}{\sum_j k_j}, \tag{3.33}$$

for some constant A. As $\delta t \to 0$

$$\frac{dk_i}{dt} = A \frac{k_i}{\sum_j k_j}. \tag{3.34}$$

Across the entire network, in one time step $\Delta k = m$ and $\sum_j k_j = 2mt$. This is an undirected graph so each edge counts twice. Putting these into (3.34):

$$\sum_i \frac{dk_i}{dt} = A \frac{\sum_i k_i}{\sum_j k_j} = m, \tag{3.35}$$

which allows us to eliminate A and leaves the differential equation for k_i as follows:

$$\frac{dk_i}{dt} = \frac{mk_i}{2mt} = \frac{k_i}{2t}. \tag{3.36}$$

This is now separable:

$$\frac{1}{k_i} \frac{dk_i}{dt} = \frac{1}{2t}, \tag{3.37}$$

with solution:

$$\ln k_i = \frac{1}{2} \ln t + \bar{C}, \tag{3.38}$$

or, neater:

$$k_i(t) = Ct^{-/2}. \tag{3.39}$$

The initial condition $k_i(t_i) = m$, which in turn allows us to eliminate C:

$$k_i(t) = m \left(\frac{t}{t_i} \right)^{1/2}. \tag{3.40}$$

If t_i is smaller, k_i is larger and older nodes are more likely to be hubs. Hence the "rich get richer".

$$p(k_i(t) < k) = p \left(t_i > \frac{m^2 t}{k^2} \right). \tag{3.41}$$

If δt is fixed, the probability density of t_i is

$$P_i(t_i) = \frac{1}{m_0 + t}. \tag{3.42}$$

Implying:

$$P \left(t_i > \frac{m^2 t}{k^2} \right) = 1 - P \left(t_i \le \frac{m^2 t}{k^2} \right) = 1 - \frac{m^2 t}{k^2 (t + m_0)}. \tag{3.43}$$

So, the probability density of $P(k)$ is as follows:

$$P(k) = \frac{\partial P(k_i(t) < k)}{\partial k} = \frac{2m^2 t}{m_0 + t} k^{-3}. \tag{3.44}$$

This is the explicit form for the cumulative degree distribution given in (3.18), hence the model of preferential attachment yields a scale-free network with $\alpha = 3$.

Although we have not included the code here, running this as a simulation is fairly straightforward and the numerical results are in good agreement with the analytical solution, giving around $\alpha = 2.9$.

3.4.2.   *Your friends are more popular than you*

We promised to come back to some of the interesting properties of the scale-free network. There is one in particular that has become known as the friendship paradox.

Let us pick one node b in a scale-free friendship network at random, and affectionately call him Billy. The probability p_k is the probability that b has k friends and takes the form given in (3.17). Let us assume that $k_{\min} = 1$ in this network of n people, which makes the expected degree of b:

$$\langle k \rangle = \sum_{k=1}^{k_{\max}} k p_k = (\alpha - 1) \sum_{k=1}^{k_{\max}} k^{1-\alpha}. \tag{3.45}$$

Now let us look at one of Billy's friends at random by picking an edge and examining the degree of the node it leads us to. Label this new node as $\hat{b}$. How many friends does Billy's friend have? Or, equivalently, what is the expected degree of $\hat{b}$?

If p_1 is the probability that a node has one edge then np_1 is the number of people with one friend. Likewise, $2np_2$ is the number of edges which link to people with two friends. And knp_k is the number of edges which link to nodes of degree k.

So, the probability of picking an edge which links to a friend with degree k is

$$\frac{knp_k}{n\langle k \rangle} = \frac{kp_k}{\langle k \rangle}. \tag{3.46}$$

This is effectively the probability that $\hat{b}$ will have degree k, or k friends, and in turn that makes the expected degree of $\hat{b}$:

$$\sum_{k=1}^{\infty} \frac{k^2 p_k}{\langle k \rangle} = \frac{\langle k^2 \rangle}{\langle k \rangle}. \tag{3.47}$$

However, we know from (3.31) that the right-hand side in (3.47) can be split, and so the expected number of connections of Billy's friend is:

$$\frac{\langle k^2 \rangle}{\langle k \rangle} = \frac{\sigma^2 + \langle k \rangle^2}{\langle k \rangle} \tag{3.48}$$

$$= \frac{\sigma^2}{\langle k \rangle} + \langle k \rangle. \tag{3.49}$$

The second term here, $\langle k \rangle$, is just the expected degree of b, i.e. how many friends we expect Billy to have. Meanwhile, the first term in (3.49) must be positive and thus the expected degree of $\hat{b}$ is higher than that of b. Billy's friend is more popular than he is.

This just demonstrates how the mean of a power law distribution conjures up some issues that one needs to be aware of. It is like saying that the vast majority of people have an above average number of arms. Which is true, but does not really represent the system.

3.5. *Epidemics on a network*

Complex network structures certainly have some interesting properties, but the most interesting examples consider how the topology affects the dynamics on the network. To illustrate, consider a model of infection spreading through a population.

Before we get to the impact of the network, let us first consider how the dynamics works for a fully mixed population of *susceptible*, *infected* and *recovered* individuals.

3.5.1. *A basic SIR model*

Assuming a fully mixed population (i.e. all individuals are connected to all others).

S The number of susceptible individuals in the population.
I The number of infected individuals in the population.
R The removed/recovered individuals who have immunity from further infection.
ν Infection rate. The probability of infection in one time step, given that an individual has had contact with an infected party.
δ Recovery rate. The probability of recovery in one time step.

Newly infected individuals will leave the S group at every time step and join the I group. Likewise, recovered individuals will leave the I group and join the R group. Meanwhile, only infected individuals can infect others. These together make the difference equations for the dynamics fairly intuitive:

$$S(t + \delta t) = S(t) - \delta t \nu I(t) S(t), \qquad (3.50)$$

$$I(t + \delta t) = I(t) + \delta t \nu I(t) S(t) - \delta I(t) \delta t, \qquad (3.51)$$

$$R(t + \delta t) = R(t) + \delta I(t) \delta t. \qquad (3.52)$$

And leads to the continuous time model:

$$\frac{dS}{dt} = -\nu I S, \quad \frac{dI}{dt} = \nu I S - \delta I, \quad \frac{dR}{dt} = \delta I \qquad (3.53)$$

Births and deaths are ignored in this model.

For a disease to die off, we want $dI/dt < 0$:

$$\frac{dI}{dt} = I(\nu S - \delta) < 0, \qquad (3.54)$$

which implies:

$$\frac{\nu S}{\delta} < 1. \qquad (3.55)$$

Thus we define:

$$R_0 = \frac{\nu S}{\delta}, \qquad (3.56)$$

as the reproductive rate or, the number of secondary infections caused by a single infected individual before it dies. We can get this another way: as the recovery rate implies the length of an illness is exponentially distributed with mean:

$$T = \frac{1}{\delta}. \tag{3.57}$$

If on average an individual can infect νS others in one time step, it can infect:

$$R_0 = \nu ST = \frac{\nu S}{\delta}, \tag{3.58}$$

before it dies.

3.5.2. *SIR model on a network*

Of course, everything changes once you take into account the fact that the population is not fully mixed. In reality, social networks tend to be scale free which means connections are not democratic. The impact of these features on the dynamics of the disease can be studied as follows:

S_k The fraction of nodes of degree k which are susceptible.
I_k The fraction of nodes of degree k which are infected.
R_k The fraction of nodes of degree k which are removed/ recovered.
ν Infection rate. The probability of infection in one time step, given that an individual has had contact with an infected party.
δ The recovery rate.

If the network was fully connected, the dynamics of the epidemic would follow that of the fully mixed case above. However, we were not considering a fully connected network.

Here, the number of edges connecting to an infected node of degree k will be nkI_k. Thus, the probability of an edge being connected to an infected node of degree k is

$$\frac{kI_k}{\langle k \rangle}. \tag{3.59}$$

So

$$\frac{dS_k}{dt} = -\nu S_k k \sum_{\hat{k}} \frac{\hat{k} I_{\hat{k}}}{\langle k \rangle}, \tag{3.60}$$

$$\frac{dI_k}{dt} = +\nu S_k k \sum_{\hat{k}} \frac{\hat{k} I_{\hat{k}}}{\langle k \rangle} - \delta I_k, \tag{3.61}$$

$$\frac{dR_k}{dt} = \delta I_k. \tag{3.62}$$

Let us put

$$\rho_0 = \frac{\nu \langle k \rangle}{\delta}, \tag{3.63}$$

which is the average number of secondary infections caused by a single infected individual in the same way as the mixed model.

But picking an infected edge, the expected degree of the attached node is

$$\sum \frac{k^2 I_k}{\langle k \rangle} = \frac{\langle k^2 \rangle}{\langle k \rangle}. \tag{3.64}$$

Now

$$R_0 = \frac{\nu}{\delta} \frac{\langle k^2 \rangle}{\langle k \rangle}. \tag{3.65}$$

Put

$$Cv^2 = \frac{\langle k^2 \rangle}{\langle k \rangle^2} - 1, \tag{3.66}$$

where Cv_2 is the coefficient of variation of the connectivity distribution. Then

$$R_0 = \rho_0(1 + Cv^2). \tag{3.67}$$

So, $\rho_0 = R_o$ when $\langle k^2 \rangle = \langle k \rangle^2$ (as in a lattice). And the critical value is $\rho_0 = 1$ or, $\nu/\delta = 1/\langle k \rangle$.

But

$$\sigma^2 = \langle k^2 \rangle - \langle k \rangle^2, \tag{3.68}$$

which suggests

$$R_0 = \frac{\nu}{\delta} \frac{\sigma^2 + \langle k \rangle^2}{\langle k \rangle} \tag{3.69}$$

$$= \frac{\nu}{\delta} \left(\frac{\sigma^2}{\langle k \rangle} + \langle k \rangle \right). \tag{3.70}$$

But in a scale-free network, $\sigma^2, \langle k \rangle \to \infty$ and the number of infections caused by a single infected individual is infinite. Viruses are unstoppable!

This example offers a simple demonstration of how the network topology can alter dynamics within a system, but the growing interest in networks and their properties has uncovered a great deal more than we could include here. For an excellent introduction, see Newman's book on the subject.[8]

3.6. *Exercises*

Exercise 3.1. For each node of the network in Fig. 6, calculate

(a) The clustering coefficient
(b) The degree centrality
(c) The closeness centrality
(d) The betweenness centrality.

Exercise 3.2. If an infected individual recovers and becomes susceptible again rather than immune, the model is known as SIS rather than SIR. If ρ_k is the probability that a node of degree k is infected,

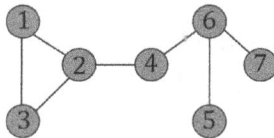

Fig. 6. G_4.

ν is the infection rate and δ the recovery rate, explain why:

$$\frac{\partial \rho_k}{\partial t} - \delta \rho_k + \nu k(1 - \rho_k)\theta, \qquad (3.71)$$

can be said to govern the dynamics of an SIS model in a scale-free network. Determine the form of θ.

4. Entropy Maximisation

Although complex networks and agent-based models differ vastly in their approaches and techniques, broadly speaking, they both allow you to consider interactions at the individual level and then scale upwards to study the system as a whole. In that sense, they both approach complexity from the same end of Weaver's spectrum of problems, through the tools of Problems of Simplicity.

In our final section, we aim to find a way into complex systems from the other end, by approximating the system as a Problem of Disorganised Complexity. Doing so allows the statistical tools of information theory to come into play, including one particularly useful method which is used in everything from machine learning to the study of urban systems: Shannon entropy maximisation.

Consider, for example, an urban system like the shopping centres of a city. The information there is contained in three hierarchies. At the micro-level you have the individual shoppers, travelling from their homes, deciding where to spend their money and choosing particular shops. At the macro-level you have some easily collectable broad statistics of the system, perhaps the total amount of money spent in each centre per day. At the meso-level, in between the other two, you might know the total spending from each residential area to each shopping centre, but not precisely where each individual chooses to go shopping.

Entropy maximisation methods allow you to translate between the different levels of the system by creating a probability distribution at the meso-level. In our retail example (as we shall see later in Section 4.3) it provides a probability distribution of the most likely spending flows between residential areas and shopping centres.

More generally, entropy maximisation helps you to average out the micro-level interactions in an unbiased way while providing a richer description of the system than is offered by the macro-level.

4.1. *The Shannon entropy*

At the centre of the method is the *Shannon entropy*, defined as:

$$S = -\sum_x P(x)\ln P(x). \tag{4.1}$$

Here x is a particular micro-state and $P(x)$ is the probability distribution of x. The objective is to find the form of P which maximises S in (4.1), subject to what is known about the system. To quote E.T Jaynes, who first described the method entropy maximisation in his paper of 1957:

"In making inferences on the basis of partial information we must use that probability distribution which has maximum entropy subject to whatever is known. This is the only unbiased assumption we can make; to use any other would amount to arbitrary assumption of information which by hypothesis we do not have"

But why maximise? And what has this got to do with bias? Rather than break with the habit of numerous mathematicians' lifetimes — let us illustrate using the example of a coin toss.

In a coin toss, there are two micro-states: heads and tails. The Shannon entropy in this case is simply:

$$S = -P(H)\ln P(H) - (1 - P(H))\ln(1 - P(H)), \tag{4.2}$$

and plotted in Fig. 7. There, S is at its maximum of $S = -\ln 2$ when the coin is perfectly fair and $P(H) = P(T) = 0.5$.

Naturally if you knew nothing about the coin and were choosing a probability distribution for it, you should select the most unbiased option: that with the highest entropy and that of a fair coin.

If you had more information about the macro-state of the system (for example, the average number of heads in 100 throws) you can incorporate this into the entropy maximisation via a series of

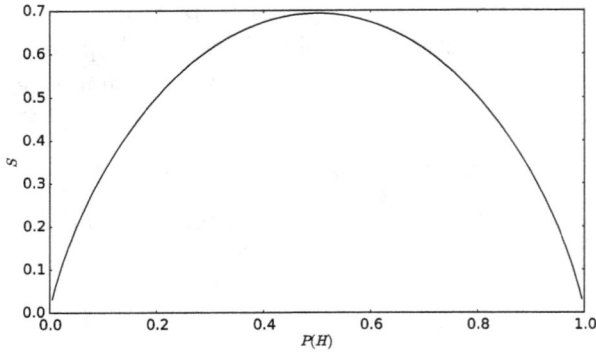

Fig. 7.   The entropy of a single coin toss.

constraints (as we shall see in Section 4.2), but the principle of the method remains the same: regardless of what is known about the system you should select the most unbiased probability distribution subject to the information available.

This might seem fairly obvious in the example of a coin, but entropy maximisation comes into its own in larger systems. Especially situations where the possible number of configurations, or micro-states, N is much greater than the number of observations of the system k. $N \gg k$.

4.2.   *Applying constraints*

Finding a $P(x)$ to maximise the Shannon entropy in (4.1) might be relatively straightforward with some basic calculus, but it is also important to include what is known about the system at the macro-level before selecting a probability distribution to represent it. How to approach the constraints is best explained through an example. We will use a simple one first in this section, and then with a real-world complex system in Section 4.3.

Imagine you are trying to decide whether to go to the nice coffee shop around the corner to pick up some breakfast. Sometimes, you will go there and be served straight away, at other times the shop will be very busy and you will have to wait to be served. Their pastries are delicious, but you have got a busy day to be getting on with. So you do what any good mathematician would do and formulate an entropy maximisation framework to estimate the probability distribution of how long you will have to wait for your breakfast.

In previous weeks, as a diligent record keeper, you have kept a note of how long you have spent waiting. In minutes, these are: $0, 0, 1, 3, 5, 5, 10, 16$ making your expected waiting time, $E(x) = 5$ min.

If you were basing $P(x)$ on your observations alone, you could claim that waiting 1 minute had $P(1) = 1/8$, or not waiting at all was $P(0) = 1/4$. However, this would create a problem when trying to estimate the probability for values you have not yet encountered, since it would imply $P(2) = P(4) = 0$.

Knowing there are more configurations possible than those you have observed, you could decide to assume the probability distribution was a Gaussian or similar and estimate the standard deviation of the underlying probability distribution based on your data, but that would be imposing unnecessary assumptions on the distribution and potentially adding bias to the system.

Let us assume we are only interested in discrete times.[a] The probability distribution $P(x)$ must meet the following constraints:

(I) $\displaystyle\sum_{x=0}^{\infty} P(x) = 1$ All probabilities must sum to one.

(II) $\displaystyle\sum_{x=0}^{\infty} xP(x) = 5$ The expected waiting time is 5 minutes.

Figure 8 illustrates the objective: we are looking for the value of P at x which gives the highest point on the entropy surface S, as in (4.1), that is also along the line of our constraints (I) and (II), represented by $c(x, P)$.

This will happen when the gradients of S and c are parallel — as illustrated by the sketch in Fig. 8(b).

---

[a] Afterall, you are doing this calculation instead of queuing for your breakfast. Those extra seconds probably are not going to make much of a difference.

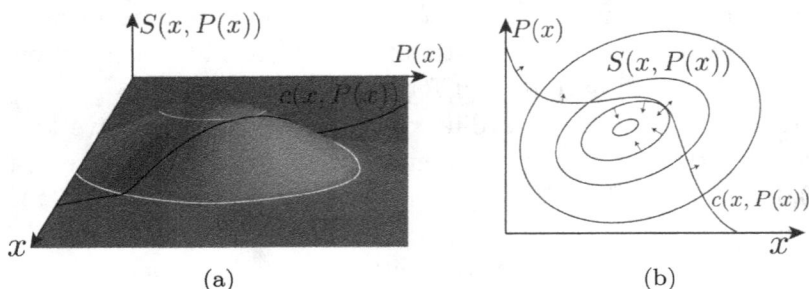

(a) (b)

Fig. 8. A sketch of the entropy surface and constraints. We seek to find the form of $P(x)$ which maximises S while satisfying c.

To find this point, we use the method of Lagrangian Multipliers, which begins by defining a new function:

$$\Lambda(P, \alpha, \beta) = -\sum_{x} P(x) \ln P(x)$$

$$-\alpha \left(\sum_{x=0}^{\infty} xP(x) - 5 \right) - \beta \left(\sum_{x=0}^{\infty} P(x) - 1 \right). \quad (4.3)$$

The task is now to find the turning point of Λ.

$$\frac{\partial \Lambda}{\partial \alpha} = 0, \quad \frac{\partial \Lambda}{\partial \beta} = 0, \quad \frac{\partial \Lambda}{\partial P} = 0. \quad (4.4)$$

The first two of these, the derivatives of Λ with respect to α and β, will fix the constraints (I) and (II). Meanwhile, the derivative with respect to P at x will give us the point where the gradients of our constraints are parallel to the gradient of S, as required.

The derivative with respect to P at x will knock out all the terms in the summation except for those at x:

$$\frac{\partial \Lambda}{\partial P} = -(\ln P(x) + 1) - \alpha x - \beta = 0. \quad (4.5)$$

We now have an expression for P in terms of x:

$$P(x) = \frac{e^{-\alpha x}}{e^{1+\beta}}. \quad (4.6)$$

Let us relabel the denominator in (4.6):

$$Z = e^{1+\beta}, \quad (4.7)$$

so that:

$$P(x) = \frac{e^{-\alpha x}}{Z}. \tag{4.8}$$

Now Z works as a normalisation constant and, by analogy with Statistical Physics, is known as the partition function. To determine our probability distribution, all that remains is to find the values of α and Z. This can be done by applying the two constraints. First, (I) implies:

$$\sum_{x=0}^{\infty} e^{-\alpha x}/Z = 1, \tag{4.9}$$

or rather,

$$Z = \sum_{x=0}^{\infty} e^{-\alpha x}, \tag{4.10}$$

while the second constraint:

$$\sum_{x=0}^{\infty} xe^{-\alpha x}/Z = 5, \tag{4.11}$$

can now be written entirely in terms of Z since, thanks to (4.10):

$$\frac{\partial Z}{\partial \alpha} = -\sum_{x=0}^{\infty} xe^{-\alpha x}. \tag{4.12}$$

The form of (4.12) then makes constraint (II) simply:

$$-\frac{1}{Z}\frac{\partial Z}{\partial \alpha} = 5. \tag{4.13}$$

And actually, the expression for (4.10) can take us further. If we expand out the infinite series for Z given there:

$$Z = 1 + e^{-\alpha} + e^{-2\alpha} + e^{-3\alpha} + \cdots, \tag{4.14}$$

it becomes clear that Z is just a standard geometric series, which means:

$$Z = \frac{1}{1 - e^{-\alpha}}. \tag{4.15}$$

Putting this into (4.13):

$$(1 - e^{-\alpha})\frac{e^\alpha}{(1 - e^{-\alpha})^2} = 5, \tag{4.16}$$

which can be rearranged to give:

$$\alpha = -\ln\left(\frac{5}{6}\right), \tag{4.17}$$

and hence the probability distribution for your waiting time:

$$P(x) = (1 - e^{\ln(5/6)})e^{\ln(5/6)x}, \tag{4.18}$$

which, itself simplifies to

$$P(x) = 5^x 6^{-x-1}. \tag{4.19}$$

A neat solution, but by now it is almost night time and the coffee shop is closed.

   This simple example illustrates the theory, but the waiting times of a coffee shop do not quite fit the criteria of a complex system. For that we switch to our next example, that of a system of shopping centres across a city.

4.3.   *Retail*

The application of entropy maximisation to a complex system of retail centres was one of the first in the field. Since then, the concept has been applied to studies of migration,[3] archaeological sites[2] and riots.[1] The particular example used here illustrates how averaging out micro-level interactions over short time scales using statistical methods can allow you to build in feedback and nonlinearity over a longer time scale, hence creating an analysable framework capable of replicating the dynamics of a real-world complex system.

   Consider a spatial region with residential areas (i) and shopping centres (j). It is relatively straightforward to determine, from data, the disposable income of a residential area, but the task here is to try and determine where the individual residents will shop, and hence the money spent by shoppers from area i in each shopping centre j at any point in time.

If we define:

X_i The total money spent by shoppers from residential area i.
Z_j The floor space of shopping centre j.
P_{ij} The probability of one unit of money flowing from i to j.
Y_{ij} The money spent in shopping centre j by residents of area i.
d_{ij} The distance between i and j.

This system has a hierarchical information structure (as outlined in Section 4):

The macro-state: We have information about the total money spent in each centre Z_j and from area X_i.

The meso-state: Below this, a more detailed picture would be given by the probability of one money unit flowing from each i and spent in each j. This second state is the distribution P_{ij}.

The micro-state: Below this, a complete description to the entire problem would be given by the spending pattern of each individual in i; where they shopped and how much they spent.

We want to use the information from the macro-level, along with some simple assumptions at the micro-level about how people behave, to find P_{ij} and hence the most probable set of spending flows, Y_{ij}. This a model that has proved to be invaluable to retailers over a number of years, for obvious reasons.

First, let us define a utility function for each shopper in i and their costs and benefits for choosing shop j. Here, we assume that people will, on average choose shops which maximise their individual utility. In the simplest form, let us take the floorspace of a shop as a benefit, but one with diminishing returns, while distance from the shopper detracts from the attractiveness:

$$U_{ij} = a \ln Z_j - b d_{ij}. \tag{4.20}$$

This has been shown to stand up well in empirical studies: People like local shopping centres, but are prepared to travel further for a really big centre.

We also need some constraints. The following constraints are traditionally chosen, partly by analogy with Statistical Physics, partly thanks to the form of the utility function, but mostly because they work to produce good empirical results.

(I) $\sum_j P_{ij} = 1$ The money spent by shoppers from i in all centres j must equal the total.

(II) $\sum_{ij} P_{ij} d_{ij} = C$ The total cost has some value C, analogous in some sense to system energy.

(III) $\sum_{ij} P_{ij} \log Z_j = B$ There is some total benefit or capacity in the system.

4.3.1. *A probability distribution for spending flow*

The task is to maximise the Shannon Entropy:

$$S = - \sum_{ij} P_{ij} \ln P_{ij}, \qquad (4.21)$$

subject to the constraints (I), (II) and (III). To do so, we again use the method of Lagrangian multipliers, i.e. by solving $\nabla \Lambda = 0$, where

$$\Lambda(P_{ij}, \alpha, \beta, \gamma) = - \sum_{ij} P_{ij} \log P_{ij} + \alpha \left(\sum_{ij} P_{ij} \log Z_j - B \right)$$

$$- \beta \left(\sum_{ij} P_{ij} d_{ij} - C \right) - \cdots - \sum_i \gamma_i \left(\sum_j P_{ij} - 1 \right), \qquad (4.22)$$

for Lagrangian multipliers α, β γ_i.

Again $\frac{\partial \Lambda}{\partial \alpha} = 0$, $\frac{\partial \Lambda}{\partial \beta} = 0$ and $\frac{\partial \Lambda}{\partial \gamma_i} = 0$ fix the constraints, and:

$$\frac{\partial \Lambda}{\partial P_{ij}} = - \log P_{ij} - 1 + \alpha \log Z_j - \beta d_{ij} - \gamma_i = 0, \qquad (4.23)$$

leads directly to the result

$$P_{ij} = \frac{Z_j^\alpha \exp(-\beta d_{ij})}{\exp(1 + \gamma_i)}. \qquad (4.24)$$

The first constraint in (I) implies:

$$1 = \sum_k \frac{Z_k^\alpha \exp\left(-\beta d_{ik}\right)}{\exp\left(1 + \gamma_i\right)}, \tag{4.25}$$

which allows us to eliminate the Lagrangian multiplier γ_i, and to rewrite (4.24) as follows:

$$P_{ij} = \frac{Z_j^\alpha \exp(-\beta d_{ij})}{\sum_k Z_k^\alpha \exp(-\beta d_{ik})}. \tag{4.26}$$

This form of P_{ij} also makes good intuitive sense. The $\exp(\alpha \log Z_j - \beta d_{ij})$ in the numerator is just the exponential of the utility function in (4.20) and speaks of the total advantage or pulling power of centre j to area i, weighted by the constants α and β. Meanwhile, the denominator acts as a competition term, comparing the attractiveness of centre j to all other centres.[b]

Finally, since X_i is the amount of money units spent by each residential area, we can use (4.26) to write down the most probable set of spending flows:

$$Y_{ij} = \frac{X_i Z_j^\alpha \exp(-\beta d_{ij})}{\sum_k Z_k^\alpha \exp(-\beta d_{ik})}. \tag{4.27}$$

The values of α and β here are still unknown, but — unlike in the example of Section 4.2 — these can be determined from the data, allowing the model to be fitted empirically to a particular retail system.

4.3.2. A dynamic model for centre size

Ignoring any interaction at the micro-scale and approximating the system of retail centres as a problem in disorganised complexity allows us to derive a probability distribution for the most likely set of spending flows at any point in time. But this does not mean we need to exclude interaction entirely from the model. Instead, as we see in this section, the form of Y_{ij} and its relationship to centre size

---

[b]The distribution of P_{ij} in (4.26) is analogous to the Boltzmann form seen in statistical mechanics. It is also equivalent to the clogit model of McFadden's discrete choice, albeit through a different derivation.

Z_j can now be embedded into a dynamic framework, allowing the flow of money into a retail centre to affect its size.

If Y_{ij} is known for a given time step, then D_j, the money taken by shopping centre j, may be found from

$$D_j = \sum_i Y_{ij} \tag{4.28}$$

since the total flow of money into j must be equivalent to the sum of the flows from each residential area i.

If we assume the costs of running a retail centre are linearly related to its size, we can write down an expression for the profit (or loss) of a retail centre:

$$\text{Profit} = D_j - kZ_j, \tag{4.29}$$

for some constant k.

Further if we assume the per unit growth per unit time of a retail centre is linearly related to its size, we can construct a difference equation for Z_j at the next time step:

$$\frac{1}{Z_j^t} \Delta Z_j^{t+1} = \delta t \left(D_j^t - kZ_j^t \right). \tag{4.30}$$

And, as $\delta t \to 0$, we can consider the continuous time case:

$$\frac{1}{Z} \frac{\partial Z_j}{\partial t} = \left(\frac{X_i Z_j^\alpha \exp(-\beta d_{ij})}{\sum_k Z_k^\alpha \exp(-\beta d_{ik})} \right) - kZ_j \right). \tag{4.31}$$

Now the growth of all retail centres are influenced by all others and we have feedback in the system.

The $1/Z_j$ term on the left-hand side governs the dynamics of growth near $Z_j = 0$ since it means small centres grow slower than large ones. It also builds in many similarities to the Lotka–Volterra systems seen in Chapter 1, and much of the analytical tools there also have a relevance here.

4.3.3.   The role of α

The parameters α and β are found from the data, but serve to govern the dynamics of the system. To demonstrate the impact of α let us make the simplifying assumption that all centres stay the same size for now. Differentiating (4.27) with respect to Z_j:

$$\frac{\partial Y_{ij}}{\partial Z_j} = \frac{X_i \alpha Z_j^{-\alpha-1} \exp(-\beta d_{ij})}{\sum_k Z_k^\alpha \exp(-\beta d_{ik})} - \frac{X_i \alpha Z_j^{\alpha-1} Z_j^\alpha \exp(-\beta d_{ij})}{\left(\sum_k Z_k^\alpha \exp(-\beta d_{ik})\right)^2}, \quad (4.32)$$

which, using (4.27) to simplify, becomes:

$$\frac{\partial Y_{ij}}{\partial Z_j} = \alpha \frac{Y_{ij}}{Z_j}\left(1 - \frac{Y_{ij}}{Z_j}\right). \quad (4.33)$$

So, as $Z_j \to \infty$, $\partial Y_{ij}/\partial Z_j \to 0$, and $\delta Y_{ij}/\partial Z_j \to 0$, suggesting that the marginal flows into a centre diminish for both very large and very small shops.

Now, assuming $\alpha > 0$, when $Z_j \ll 1$, $Y_{ij} \ll 1$ and

$$\frac{\partial Y_{ij}}{\partial Z_j} \simeq \alpha \frac{Y_{ij}}{Z_j}. \quad (4.34)$$

Summing (4.34) over i suggests:

$$\frac{\partial D_j}{\partial Z_j} \simeq \alpha Z_j^{\alpha-1} \sum_i \frac{X_i \exp(-\beta d_{ij})}{\sum_k Z_k^\alpha \exp(-\beta d_{ik})}, \quad (4.35)$$

and:

$$\frac{\partial D_j}{\partial Z_j} = \begin{cases} 0 & \text{if } \alpha > 1, \\ \text{finite} & \text{if } \alpha = 1, \\ \infty & \text{if } \alpha < 1. \end{cases} \quad (4.36)$$

Thus, small centres will grow when $\alpha < 1$ and die off when $\alpha < 1$.

More on the role of α and other analysis of the system can be found in Refs. 4 and 5.

4.4. *Exercises*

Exercise 4.1. In a system with three possible micro-states $x = 1, 2, 3$ and mean $E(x) = 2$, find probabilities $p(1), p(2), p(3)$ which maximise the entropy of the system.

Exercise 4.2. By applying the transformation:

$$Z_j^t = \zeta_j^t \left(\frac{1 + \delta t D_j}{\delta t k} \right), \tag{4.37}$$

show that the retail model given in (4.30) is equivalent to the logistic map:

$$\zeta_j^{t+\delta t} = a\zeta_j^t \left(1 - \zeta_j^t\right) \tag{4.38}$$

and determine the value of a.

5.   Solutions

Exercise 3.1 Solution.

Node	Clustering Coefficient	Degree Centrality	Closeness Centrality	Betweenness Centrality
1	1	1/3	6/15	0
2	1/3	1/2	6/11	8/15
3	1	1/3	6/15	0
4	0	1/3	6/10	9/15
5	0	1/6	6/16	0
6	0	1/2	6/11	9/15
7	0	1/6	6/16	0

Exercise 3.2 Solution. If ρ_k is the probability that a node of degree k is infected, $(1 - \rho_k)$ will be the probability a node of degree k is susceptible.

The number of edges connecting to a node of degree k is $nk\rho_k$ and hence the probability that a neighbour is degree k is

$$\frac{kp_k}{\langle k \rangle}. \tag{5.1}$$

Thus the probability that a link connects to an infected node is

$$\sum_k \frac{kp_k}{\langle k \rangle} \rho_k. \tag{5.2}$$

At each time step, $-\delta\rho_k$ nodes will recover, while $(1-\rho_k)k$ links from susceptible nodes may be connected to an infected individual. This leads directly to the differential equation for ρ_k:

$$\frac{\partial \rho_k}{\partial t} = -\delta \rho_k + \nu(1 - \rho_k)k \sum_k \frac{kp_k}{\langle k \rangle} \rho_k. \tag{5.3}$$

And thus:

$$\theta = \sum_k \frac{kp_k}{\langle k \rangle} \rho_k. \tag{5.4}$$

For more on SIS dynamics on a scale free network see Ref. 11.

Exercise 4.1 Solution. The constraints are:

$$(I) \quad \sum_{x=1}^{x=3} p(x) = 1 \qquad (II) \quad \sum_{x=1}^{3} xp(x) = 2.$$

We wish to maximise $S = -\sum_{x=1}^{3} p(x) \ln(p(x))$. We do so by using the Lagrangian:

$$\Lambda = -\sum_{x=1}^{3} p(x) \ln(p(x)) - \alpha \left(\sum_{x=1}^{3} xp(x) - 2 \right) - \beta \left(\sum_{x=1}^{x=3} p(x) - 1 \right). \tag{5.5}$$

Then $\partial \Lambda / \partial \alpha = 0$ and $\partial \Lambda / \partial \beta = 0$ fix the constraints, while $\partial \Lambda / \partial p(x) = 0$ imply:

$$p(x) = \frac{e^{-\alpha x}}{Z}, \tag{5.6}$$

as in the example in Section 4.2. Here, the first constraint implies:

$$Z = e^{-\alpha} + e^{-2\alpha} + e^{-3\alpha}, \tag{5.7}$$

and so:

$$p(x) = \frac{e^{-\alpha x}}{e^{-\alpha} + e^{-2\alpha} + e^{-3\alpha}} = \frac{e^{(1-x)\alpha}}{1 + e^{-\alpha} + e^{-2\alpha}}. \tag{5.8}$$

The second constraint then implies:

$$\sum_{x=1}^{3} xp(x) = \frac{e^{0\alpha} + 2e^{-\alpha} + 3e^{-2\alpha}}{1 + e^{-\alpha} + e^{-2\alpha}} = 1 + \frac{e^{-\alpha} + 2e^{-2\alpha}}{1 + e^{-\alpha} + e^{-2\alpha}} = 2, \tag{5.9}$$

which simplifies to:

$$e^{-2\alpha} = \left(e^{-\alpha}\right)^2 = 1. \tag{5.10}$$

Taking the positive square root: $e^{-\alpha} = 1$ gives the following result for $p(1), p(2), p(3)$:

$$p(1) = \frac{e^0}{1 + 1 + 1^2} = \frac{1}{3}, \ \ p(2) = \frac{1}{3}, \ \ p(3) = \frac{1}{3}. \tag{5.11}$$

Exercise 4.2 Solution: Substituting $Z_j^t = \zeta_j^t\left(\frac{1+\delta t D_j}{\delta t k}\right)$ into (4.30) and rearranging gives:

$$\zeta_j^{t+\delta t} = (1 + \delta t D_j)\,\zeta_j^t\left(1 - \zeta_j^t\right). \tag{5.12}$$

References

1. T. P. Davies, H. M. Fry, A. G. Wilson and S. R. Bishop, A mathematical model of the london riots and their policing, *Sci. Rep.* **3**, 1303 (2013).
2. T. P. Davies, H. M. Fry, A. G. Wilson, A. Palmisano, M. Altaweel and K. Radner, Application of an entropy maximizing and dynamics model for understanding settlement structure: the Khabur triangle in the middle bronze and iron ages, *J. Archaeological Sci.* **43**, 141–154 (2014).
3. A. Dennett and A. G. Wilson, A multilevel spatial interaction modelling framework for estimating interregional migration in Europe, *Environ. Plan. A* **45**(6), 1491–1507 (2013).
4. H. M. Fry and F. T. Smith, Rate effects on the growth of centres, *European J. Appl. Math.* **First View**, 1–22 (2016).
5. B. Harris and A. G. Wilson, Equilibrium values and dynamics of attractiveness terms in production-constrained spatial-interaction models, *Environ. Plan. A* **10**(4), 371–388 (1978).
6. J. Ladyman, J. Lambert and K. Wiesner, What is a complex system? *Euro. Jnl. Phil. Sci.* **3**(1), 33–67 (2013).
7. M. E. J. Newman, Complex systems: A survey, *Am. J. Phys.* 79(8), 800 (2011).
8. M. E. J. Newman, *Networks: An introduction*, Oxford University Press (2010).
9. M. Niazi and A. Hussain, Agent-based computing from multi-agent systems to agent-based models: A visual survey, *Scientometrics* **89**, 479–499 (2011).
10. T. Oleron Evans, Perspectives on the relationship between local interactions and global outcomes in spatially explicit models of systems of interacting individuals, Ph.D. Thesis (2015).
11. R. Pastor-Satorras and A. Vespignani. Epidemic spreading in scale-free networks, *Phys. Rev. Lett.* **86**(14), 3200–3203 (2001).

Chapter 4

Dynamical Systems in Cosmology

Christian G. Böhmer

Department of Mathematics, University College London
Gower Street, London, WC1E 6BT, UK
c.boehmer@ucl.ac.uk

Nyein Chan

Faculty of Engineering, Computing and Science
Swinburne University of Technology, Sarawak Campus
Jalan Simpang Tiga, 93350 Kuching, Sarawak, Malaysia
The International School Yangon
Shwe Taungyar Street, Yangon, Myanmar
nchan@isyedu.org

Cosmology is a well-established research area in physics while dynamical systems are well established in mathematics. It turns out that dynamical system techniques are very well suited to study many aspects of cosmology. The aim of this chapter is to provide the reader with a concise introduction to both cosmology and dynamical systems. The material is self-contained with references to more detailed work. It is aimed at applied mathematics and theoretical physics graduate level students who have an interest in this exciting topic.

1. A Brief Introduction to Cosmology

1.1. *The basics*

Cosmology is the study of the universe as a whole, and its aim is to understand the origin of the universe and its evolution. The study of the cosmos is as old as humanity and has always been

fascinating. Physical cosmology[a] is the scientific study of the universe
as a whole based on the laws of physics. The dominant interaction
between macroscopic objects is the gravitational force. Therefore,
we must study the dynamics of the universe within the framework
of Einstein's theory of General Relativity which was formulated in
1916. In simple terms, the main concept of general relativity is the
following equation:

$$\text{geometry} = \kappa \times \text{matter}, \tag{1.1}$$

where $\kappa = 8\pi G/c^4$ is a coupling constant which determines the
strength of the gravitational force. G is Newton's gravitational constant and c is the speed of light.

The Einstein field equations are a set of 10 coupled nonlinear
PDEs, or in other words, very difficult equations to deal with in
general. However, these equations can be simplified considerably by
making some suitable assumptions. In cosmology,[1] this is known as
the cosmological principle. It is an axiom which states that the universe is homogeneous and isotropic when viewed over large enough
scales.

These scales are of the order of 100–1000 MPc. To translate this
into more practical units, we note that $1\,\text{pc} \approx 3.26\,\text{ly} \approx 10^{12}\,\text{km}$. This
means $100\,\text{MPc} \approx 326\,\text{Mly} \approx 10^{18}\,\text{km}$, and has the simple practical
implication that we cannot test the cosmological principle directly by
making observations at two points in the universe separated by cos-
mologically significant scales. However, there are other possibilities of
testing the cosmological principle. For instance, if we were to observe
a very large structure in the universe which is bigger than 100 MPc,
say, then this would force us to revise this number upwards. It turns
out that such very large structures have already been observed, see
the Clowes–Campusano Large Quasar Group for one such example.

Henceforth we assume that the cosmological principle is valid
for some suitable length scale. A homogeneous and isotropic
four-dimensional Lorentzian manifold is characterised by only one

---

[a]We will drop the word physical soon. It is used here to emphasise the scientific aspect
of cosmology opposed to the philosophical or religious studies.

function which is usually denoted by $a(t)$ and one constant $k = (\pm 1, 0)$. Such models were studied independently by Friedmann, Lemaître, and Robertson & Walker. The function $a(t)$ is called the scale factor and is the only dynamical degree of freedom in the cosmological Einstein field equations. The constant k characterises the curvature of the so-called constant time hypersurfaces, $k = 0$ corresponds to a Euclidean space, $k = +1$ to a 3-sphere and $k = -1$ to hyperbolic space. The cosmological Einstein field equations are given by

$$3\frac{\dot{a}^2}{a^2} + 3\frac{k}{a^2} - \Lambda = \kappa\,\rho, \tag{1.2a}$$

$$-2\frac{\ddot{a}}{a} - \frac{\dot{a}^2}{a^2} - \frac{k}{a^2} + \Lambda = \kappa\,p. \tag{1.2b}$$

Here Λ is the so-called cosmological constant, ρ and p are the energy density and pressure of some matter components, respectively. This matter could be a perfect fluid with prescribed equation of state, or a scalar field for instance. More complicated forms of matter can also be included. One can verify by direct calculation that these two equations imply the energy-conservation equation

$$\dot{\rho} + 3\frac{\dot{a}}{a}(\rho + p) = 0. \tag{1.3}$$

In cosmology, one assumes that every matter component satisfies its own conservation equation, which does not follow from the field equations but must be assumed or derived separately. Inspection of Eqs. (1.2) shows that we have two equations but three functions to be found, namely $a(t)$, $\rho(t)$ and $p(t)$. This system of equations is underdetermined. In order to close it, we will assume a linear equation of state between the pressure and the energy density $p = w\rho$, where the equation of state parameter $w \in (-1, 1]$.

The scale factor $a(t)$ is a measure of the size of the universe at time t. However, since we do not have an absolute length scale, the numerical value of $a(t)$ can be rescaled. One convention is to choose $a(t_{\text{today}}) = 1$ and compare the universe's size with its current value. Moreover, it turns out to be useful to introduce the Hubble function

$H(t) := \dot{a}/a$ which is a measure of the universe's expansion rate at time t. A positive value for this quantity was first observed by Edwin Hubble in 1929, thereby giving experimental evidence to an expanding universe. Today's value H_{today} is of the order of $70\,\text{km/s/Mpc}$.

Let us now rewrite the field equations using the Hubble parameter. Firstly, we need the relation

$$\dot{H} = \frac{\ddot{a}}{a} - \frac{\dot{a}^2}{a^2} = \frac{\ddot{a}}{a} - H^2, \tag{1.4}$$

which allows us to write (1.2) in the following form:

$$3H^2 + 3\frac{k}{a^2} - \Lambda = \kappa\,\rho, \tag{1.5a}$$

$$-2\dot{H} - 3H^2 - \frac{k}{a^2} + \Lambda = \kappa\,p. \tag{1.5b}$$

Equation (1.5a) is of particular interest to us. By dividing the entire equation by $3H^2$ we arrive at

$$1 = \frac{\kappa\,\rho}{3H^2} + \frac{\Lambda}{3H^2} - \frac{k}{a^2 H^2} \tag{1.6}$$

and observe that each of the three terms is dimensionless.

It is common to introduce the following dimensionless density parameters

$$\Omega = \frac{\kappa\,\rho}{3H^2}, \quad \Omega_\Lambda = \frac{\Lambda}{3H^2}. \tag{1.7}$$

Note that Ω may contain different forms of matter, the total matter content might contain a pressureless perfect fluid (standard matter or sometimes called dust) and radiation, in which case one would write $\Omega = \Omega_m + \Omega_r$. Before getting started with dynamical systems and their application to cosmology, we need to discuss some of the well-known solutions in cosmology.

1.2. Cosmological solutions

We will now discuss the most important solutions of the field equations (1.2). This is needed in order to understand and interpret the solutions encountered later using dynamical systems techniques.

In order to simplify the equations, we will assume that the spatial curvature parameter vanishes, i.e. $k = 0$ and we will also neglect the cosmological term $\Lambda = 0$. Let us firstly assume that the equation of state parameter $w = 0$. This corresponds to a matter dominated universe. One can immediately integrate the conservation equation (1.3) and finds that

$$\rho \propto a^{-3}. \tag{1.8}$$

This result is not unexpected since we find that density is inversely proportional to volume. Using this result in the field equation (1.2a) yields the solutions $a(t) \propto t^{2/3}$.

Secondly, we consider $w = 1/3$ which corresponds to a radiation dominated universe. In that case, the conservation equation gives

$$\rho \propto a^{-4}, \tag{1.9}$$

and the remaining field equations can be solved to find $a(t) \propto t^{1/2}$.

Lastly, we consider the case where $\rho = p = 0$, however, we assume $\Lambda > 0$. Then, we can integrate (1.5a) and find

$$a(t) \propto \exp(\sqrt{\Lambda/3}\, t). \tag{1.10}$$

This solution is generally called the de Sitter solution and corresponds to a universe which undergoes an accelerated expansion.

1.3. *A very brief history of the universe*

Based on a variety of observations, the evolution of the universe can be reconstructed fairly accurately. We are currently living in a matter dominated universe $w = 0$, and there is strong evidence for the presence of a positive cosmological constant $\Lambda > 0$. Moreover, the spatial curvature of the universe appears to be zero $k = 0$. There are some highly restrictive conditions in $k \neq 0$ models.

Since the universe is currently expanding, it must have been smaller and denser in the past. From Eqs. (1.8) and (1.9) we see that radiation decays faster than matter in an expanding universe. Therefore, at some point in the past, the universe was dominated by radiation. Going back in time further, the universe was very dense and therefore hot and relatively small. The 'beginning' of the universe is often referred to as the big bang, giving the image of a vast explosion from which the evolution of the universe started.

It appears very likely that the universe also underwent a period of accelerated expansion at its very early stages, similar to the late-time acceleration due to the cosmological term. The reasons for this are beyond the scope of this short introduction, however, we note that this epoch is called inflation.

Very roughly speaking, the standard model of cosmology can be summarised by the succession of the following eras:

$$\text{inflation} \longrightarrow \text{radiation} \longrightarrow \text{matter} \longrightarrow \text{cosmological term} \quad (1.11)$$

and a good cosmological model should be able to reproduce (parts of) this pattern.

1.4. *A first taste of dynamical systems*

In order to get a first taste of the usefulness of dynamical systems techniques in cosmology,[2,3] let us consider a universe which is spatially flat $k = 0$, and its matter content is radiation ρ_r with $w = 1/3$, and a perfect fluid (dust) ρ_m with $w = 0$. The following four equations completely determine the dynamics of the system

$$3H^2 - \Lambda = \kappa\left(\rho_m + \rho_r\right), \quad (1.12a)$$

$$-2\dot{H} - 3H^2 + \Lambda = \kappa\tfrac{1}{3}\rho_r, \quad (1.12b)$$

$$\dot{\rho}_r + 4H\rho_r = 0, \quad (1.12c)$$

$$\dot{\rho}_m + 3H\rho_m = 0. \quad (1.12d)$$

Using the dimensionless density parameters Ω_m, Ω_r and Ω_Λ, we find that Eq. (1.12a) becomes the constraint

$$1 = \Omega_m + \Omega_r + \Omega_\Lambda \quad (1.13)$$

which means that we have two independent quantities, and choose to work with Ω_m and Ω_r. Moreover, since we expect energy densities to be positive we also have the conditions $0 \leq \Omega_m \leq 1$ and $0 \leq \Omega_r \leq 1$. Therefore, also $\Omega_\Lambda \leq 1$ is needed to satisfy Eq. (1.13).

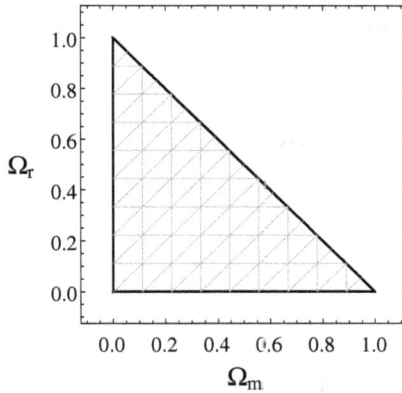

Fig. 1. This figure shows the triangle defined by $\{(\Omega_m, \Omega_r) \,|\, 0 \leq \Omega_m + \Omega_r \leq 1 \cap 0 \leq \Omega_m \leq 1 \cap 0 \leq \Omega_r \leq 1\}$. Every solution to the field equations (1.12) corresponds to a trajectory inside this triangle, one calls this region the phase space of the system.

The solution to the system (1.12) at any given time t will correspond to a point in the (Ω_m, Ω_r) plane. The constraint equation (1.13) together with the aforementioned inequalities reduces the allowed (Ω_m, Ω_r) plane to the triangle[b] defined by $\Delta = \{(\Omega_m, \Omega_r)\,|\,0 \leq \Omega_m + \Omega_r \leq 1 \cap 0 \leq \Omega_m \leq 1 \cap 0 \leq \Omega_r \leq 1\}$, see also Fig. 1.

Next, we wish to find the dynamical equations for the dimensionless variables Ω_m and Ω_r. This requires a slightly lengthy but otherwise straightforward calculation of which we will show some details. We start with

$$\frac{d}{dt}\Omega_m = \frac{d}{dt}\left(\frac{\kappa\,\rho_m}{3H^2}\right) = \frac{\kappa}{3}\frac{\dot{\rho}_m H^2 - \rho_m 2 H \dot{H}}{H^4} = \frac{\kappa}{3H}\left(\frac{\dot{\rho}_m}{H} - 2\rho_m\frac{\dot{H}}{H^2}\right).$$

$$(1.14)$$

From (1.12d) we get an expression for $\dot{\rho}_m/H$, while (1.12b) can be solved for $\dot{H}/H^2$. This yields

$$\frac{1}{H}\frac{d}{dt}\Omega_m = \frac{\kappa}{3H^2}\left(-3\rho_m + 3\rho_m\left(1 - \frac{\Lambda}{3H^2} + \frac{\kappa\,\rho_r}{9H^2}\right)\right) \qquad (1.15)$$

$$= -3\Omega_m + 3\Omega_m(1 - \Omega_\Lambda + \Omega_r/3). \qquad (1.16)$$

---

[b]To the best of our knowledge this idea goes back to Nicola Tamanini.

The last step is to eliminate Ω_Λ using (1.13) which gives the equation

$$\frac{1}{H}\frac{d}{dt}\Omega_\mathrm{m} = -3\Omega_\mathrm{m} + 3\Omega_\mathrm{m}(\Omega_\mathrm{m} + \Omega_\mathrm{r} + \Omega_\mathrm{r}/3) \qquad (1.17)$$

$$= -3\Omega_\mathrm{m} + 3\Omega_\mathrm{m}(\Omega_\mathrm{m} + 4\Omega_\mathrm{r}/3) \qquad (1.18)$$

$$= \Omega_\mathrm{m}(3\Omega_\mathrm{m} + 4\Omega_\mathrm{r} - 3). \qquad (1.19)$$

We now note that

$$d\log(a) = \frac{\dot{a}}{a}dt = H\,dt, \qquad (1.20)$$

which means that by introducing the new independent variable $N = \log(a)$ and denoting differentiation with respect to N by a prime, we finally arrive at

$$\Omega_\mathrm{m}' = \Omega_\mathrm{m}(3\Omega_\mathrm{m} + 4\Omega_\mathrm{r} - 3). \qquad (1.21)$$

Following a similar calculation one can find the corresponding equation for Ω_r which is given by

$$\Omega_\mathrm{r}' = \Omega_\mathrm{r}(3\Omega_\mathrm{m} + 4\Omega_\mathrm{r} - 4). \qquad (1.22)$$

For any set of initial conditions $(\Omega_\mathrm{m}(N_i), \Omega_\mathrm{r}(N_i))$ with initial 'time' N_i in the triangle Δ, Eqs. (1.21) and (1.22) will determine a trajectory which describes the dynamical behaviour of the cosmological model we are studying. It should be noted that Eqs. (1.21) and (1.22) do not depend explicitly on the 'time' parameter N. Such a system is called an autonomous system of equations, or a dynamical system. Equations of this type can be studied using particular methods developed for such systems. In the next section we will give a brief introduction to dynamical systems and the most common methods used to analyse them.

Exercise 1.1. Solve the cosmological field equations (1.2a) and (1.3) assuming $\Lambda = 0$, $k = +1$, and $w = 0$.

Exercise 1.2. Derive Eq. (1.22) following the derivation which led to Eq. (1.21).

2. Some Aspects of Dynamical Systems

What is a dynamical system? It can be anything ranging from something as simple as a single pendulum to something as complex as the human brain and the entire universe itself. In general, a dynamical system can be thought of as any abstract system consisting of

(1) a space (state space or phase space), and
(2) a mathematical rule describing the evolution of any point in that space.

The second point is crucial. Finding a mathematical rule which, for instance, describes the evolution of information at any neuron in the human brain is probably impossible. So, we need a mathematical rule as an input and finding one might be very difficult indeed.

The state of the system we are interested in is described by a set of quantities which are considered important about the system, and the state space is the set of all possible values of these quantities. In the case of the pendulum, the position of the mass and its momentum are natural quantities to specify the state of the system. For more complicated systems like the universe as a whole, the choice of good quantities is not at all obvious and it turns out to be useful to choose convenient variables. It is possible to analyse the same dynamical system with different sets of variables, either of which might be more suitable to a particular question.

There are two main types of dynamical systems: The first ones are continuous dynamical systems whose evolution is defined by a set of ordinary differential equations (ODEs) and the other ones are called time-discrete dynamical systems which are defined by a map or difference equations. In the context of cosmology we are studying the Einstein field equations which for a homogeneous and isotropic space result in a system of ODEs. Thus we are only interested in continuous dynamical systems and will not discuss time-discrete dynamical systems in the remainder.

Let us denote $\mathbf{x} = (x_1, x_2, \ldots, x_n) \in X$ to be an element of the state space $X \subseteq \mathbb{R}^n$. The standard form of a dynamical system is usually expressed as[4]

$$\dot{\mathbf{x}} = \mathbf{f}(\mathbf{x}), \tag{2.1}$$

where the function $\mathbf{f} : X \to X$ and where the dot denotes differentiation with respect to some suitable time parameter. We view the function $\mathbf{f}$ as a vector field on $\mathbb{R}^n$ such that

$$\mathbf{f}(\mathbf{x}) = (f_1(\mathbf{x}), \dots, f_n(\mathbf{x})). \tag{2.2}$$

The ODEs (2.1) define the vector fields of the system. At any point $x \in X$ and any particular time t, $\mathbf{f}(\mathbf{x})$ defines a vector field in $\mathbb{R}^n$. When discussing a particular solution to (2.1) this will often be denoted by $\psi(t)$ to simplify the notation. We restrict ourselves to systems which are finite dimensional and continuous. In fact, we will require the function f to be at least differentiable in X.

Definition 2.1 (Critical point or fixed point). The autonomous equation $\dot{\mathbf{x}} = \mathbf{f}(\mathbf{x})$ is said to have a critical point or fixed point at $\mathbf{x} = \mathbf{x}_0$ if and only if $\mathbf{f}(\mathbf{x}_0) = 0$.

There is an easy way to justify this definition. Let us consider a one-dimensional mechanical system with force F. Newton's equation for such a system is

$$m\ddot{x} = F(x). \tag{2.3}$$

Let us introduce a second variable $p = m\dot{x}$ such that the single second-order ODE (2.3) becomes a system of two first-order equations

$$\dot{x} = p/m, \tag{2.4}$$

$$\dot{p} = F(x). \tag{2.5}$$

Therefore, according to Definition 2.1, the critical points of system (2.5) correspond to those points x where the force vanishes $F(x) = 0$. At these points, there is no force acting on the particle and the system could, in principle, remain in this (steady) state indefinitely.

This leads to the question of stability of a critical point or fixed point. The following two definitions will clarify what is meant by stable and asymptotically stable. In simple words a fixed point x_0 of the system (2.1) is called stable if all solutions $\mathbf{x}(t)$ starting near $\mathbf{x}_0$ stay close to it.

Definition 2.2 (Stable fixed point). Let $\mathbf{x}_0$ be a fixed point of system (2.1). It is called stable if for every $\varepsilon > 0$ we can find a δ such that if $\psi(t)$ is any solution of (2.1) satisfying $\|\psi(t_0) - \mathbf{x}_0\| < \delta$, then the solution $\psi(t)$ exists for all $t \geq t_0$ and it will satisfy $\|\psi(t) - \mathbf{x}_0\| < \varepsilon$ for all $t \geq t_0$.

The point is called asymptotically stable if it is stable and the solutions approach the critical point for all nearby initial conditions.

Definition 2.3 (Asymptotically stable fixed point). Let $\mathbf{x}_0$ be a stable fixed point of system (2.1). It is called asymptotically stable if there exists a number δ such that if $\psi(t)$ is any solution of (2.1) satisfying $\|\psi(t_0) - \mathbf{x}_0\| < \delta$, then $\lim_{t \to \infty} \psi(t) = \mathbf{x}_0$.

The main difference is simply that all trajectories near an asymptotically stable fixed point will eventually reach that point while trajectories near a stable point could for instance circle around that point. If the point is unstable then solutions will move away from it.

We will not encounter fixed points which are stable but not asymptotically stable when studying cosmological dynamical systems.

Having defined a concept of stability, we will now discuss methods which can be used to analyse the stability properties of critical points.

2.1. *Linear stability theory*

The basic idea of linear stability theory can be explained neatly using the above one-dimensional mechanical system $m\ddot{x} = F(x)$. Let us assume that there is a point x_0 where the force vanishes $F(x_0) = 0$. Can we find the behaviour of the particle near this point? We set $x(t) = x_0 + \delta x(t)$ and assume $\delta x(t)$ to be small. Then $\ddot{x}(t) = \ddot{\delta x}(t)$ and $F(x) = F(x_0 + \delta x) \approx F(x_0) + F'(x_0)\delta x + \cdots = F'(x_0)\delta x + \cdots$ (recall $F(x_0) = 0$) so that Newton's equation near the critical point becomes $m\ddot{\delta x} = F'(x_0)\delta x$ where $F'(x_0)$ is a constant. This is a linear second-order constant coefficient ODE, its auxiliary equation is simply $\lambda^2 = F'(x_0)/m$. Therefore, the sign of $F'(x_0)$ determines the stability properties of the point x_0. If $F'(x_0) < 0$ the solution involves trigonometric functions and we would speak of a stable point, for $F'(x_0) > 0$ the solution would involve exponentials and we would refer to this point as unstable.

Exactly the same ideas can be utilised when studying an arbitrary dynamical system. Let $\dot{\mathbf{x}} = \mathbf{f}(\mathbf{x})$ be a given dynamical system with fixed point at $\mathbf{x}_0$. We will now linearise the system around its critical point. Since $\mathbf{f}(\mathbf{x}) = (f_1(\mathbf{x}), \ldots, f_n(\mathbf{x}))$, we can Taylor expand each $f_i(x_1, x_2, \ldots, x_n)$ near $\mathbf{x}_0$

$$f_i(\mathbf{x}) = f_i(\mathbf{x}_0) + \sum_{j=1}^{n} \frac{\partial f_i}{\partial x_j}(\mathbf{x}_0)y_j + \frac{1}{2!} \sum_{j,k=1}^{n} \frac{\partial^2 f_i}{\partial x_j \partial x_k}(\mathbf{x}_0)y_j y_k + \cdots,$$

(2.6)

where the vector $\mathbf{y}$ is defined by $\mathbf{y} = \mathbf{x} - \mathbf{x}_0$. Note that in what follows we are only interested in the first partial derivatives. Therefore, of particular importance is the object $\partial f_i/\partial x_j$ which, if interpreted as a matrix, is the Jacobian matrix of vector calculus of the vector-valued function $\mathbf{f}$. We define

$$J = \frac{\partial f_i}{\partial x_j} = \begin{pmatrix} \dfrac{\partial f_1}{\partial x_1} & \cdots & \dfrac{\partial f_1}{\partial x_n} \\ \vdots & \ddots & \vdots \\ \dfrac{\partial f_n}{\partial x_1} & \cdots & \dfrac{\partial f_n}{\partial x_n} \end{pmatrix}.$$

(2.7)

It is the eigenvalues of the Jacobian matrix J, evaluated at the critical points $\mathbf{x}_0$, which contain the information about stability. In this context J is sometimes referred to as the stability matrix of the system. As J is an $n \times n$ matrix, it will have n, possibly complex, eigenvalues (counting repeated eigenvalues accordingly). Recalling the example of the one-dimensional mechanical system at the beginning, it is clear that this approach might encounter problems if one or more of the eigenvalues are zero. This motivates the following definition.[4]

Definition 2.4 (Hyperbolic point). Let $\mathbf{x} = \mathbf{x}_0 \in X \subset \mathbb{R}^n$ be a fixed point (critical point) of the system $\dot{\mathbf{x}} = \mathbf{f}(\mathbf{x})$. Then x_0 is said to be hyperbolic if none of the eigenvalues of the Jacobian matrix $J(\mathbf{x}_0)$ have zero real part. Otherwise the point is called non-hyperbolic.

Linear stability theory fails for non-hyperbolic points and other methods have to be employed to study the stability properties.

Roughly speaking we are distinguishing three broad cases: If all eigenvalues have negative real parts, then we can regard the point as stable. If at least one of the eigenvalues has a positive real part, then the corresponding fixed point would not be stable and correspond to a saddle point which attracts trajectories in some directions but repels them along others. Lastly, all eigenvalues could have a positive real part, in which case all trajectories would be repelled.

In more than three dimensions it becomes very difficult to classify all possible critical points based on their eigenvalues. However, in two and three dimensions this can be done. In the following we present all possible cases for two-dimensional autonomous systems.

Let us consider the two-dimensional autonomous system given by

$$\dot{x} = f(x, y), \tag{2.8a}$$

$$\dot{y} = g(x, y), \tag{2.8b}$$

where f and g are (smooth) functions of x and y. We assume that there exists a hyperbolic critical point at (x_0, y_0) so that $f(x_0, y_0) = 0$ and $g(x_0, y_0) = 0$. The Jacobian matrix of the system is given by

$$J = \begin{pmatrix} f_{,x} & f_{,y} \\ g_{,x} & g_{,y} \end{pmatrix}, \tag{2.9}$$

where the $f_{,x}$ means differentiation with respect to x. Its two eigenvalues $\lambda_{1,2}$ are given by

$$\lambda_1 = \frac{1}{2}(f_{,x} + g_{,y}) + \frac{1}{2}\sqrt{(f_{,x} - g_{,y})^2 + 4f_{,y}g_{,x}}, \tag{2.10a}$$

$$\lambda_2 = \frac{1}{2}(f_{,x} + g_{,y}) - \frac{1}{2}\sqrt{(f_{,x} - g_{,y})^2 + 4f_{,y}g_{,x}}, \tag{2.10b}$$

and be evaluated at any fixed point (x_0, y_0).

Table 1. Stability or instability properties of the critical point (x_0, y_0) based on the two eigenvalues λ_1 and λ_2.

Eigenvalues	Description
$\lambda_1 < 0, \lambda_2 < 0$	The fixed point is asymptotically stable and trajectories starting near that point will approach that point $\lim_{t \to \infty}(x(t), y(t)) = (x_0, y_0)$
$\lambda_1 > 0, \lambda_2 > 0$	The fixed point is unstable and trajectories will be repelled from the point $\lim_{t \to -\infty}(x(t), y(t)) = (x_0, y_0)$. We can speak of (x_0, y_0) as the past-time attractor
$\lambda_1 < 0, \lambda_2 > 0$	The fixed point is a saddle point. Some trajectories will be repelled, others will be attracted
$\lambda_1 = 0, \lambda_2 > 0$	The point is unstable. The positive eigenvalues ensure that there is at least one unstable direction
$\lambda_1 = 0, \lambda_2 < 0$	Linear stability theory fails to determine stability. The point is non-hyperbolic and other methods are needed to study the behaviour of trajectories near that point
$\lambda_1 = \alpha + i\beta, \lambda_2 = \alpha - i\beta$	With $\alpha > 0$ and $\beta \neq 0$ the fixed point is an unstable spiral
$\lambda_1 = \alpha + i\beta, \lambda_2 = \alpha - i\beta$	With $\alpha < 0$ and $\beta \neq 0$ the fixed point is a stable spiral
$\lambda_1 = i\beta, \lambda_2 = -i\beta$	Solutions are oscillatory and the point is called a centre. Note that a critical point being a centre is not related to centre manifolds

Table 1 contains all possible cases in order to understand the stability or instability properties of the critical point (x_0, y_0) based on the two eigenvalues λ_1 and λ_2.

Example — cosmology with matter, radiation and cosmological term

Recall the cosmological dynamical system (1.21) and (1.22) which will be our base model henceforth. The equations read

$$\Omega'_m = \Omega_m(3\Omega_m + 4\Omega_r - 3), \tag{2.11a}$$

$$\Omega'_r = \Omega_r(3\Omega_m + 4\Omega_r - 4), \tag{2.11b}$$

$$1 = \Omega_m + \Omega_r + \Omega_\Lambda. \tag{2.11c}$$

We can find the fixed points of this system by solving the simultaneous equations $\Omega_{\mathrm{m}}' = 0$ and $\Omega_{\mathrm{r}}' = 0$ for the pair $(\Omega_{\mathrm{m}}, \Omega_{\mathrm{r}})$. We find three fixed points, namely $O = (0,0)$, $R = (0,1)$ and $M = (1,0)$. As we use the relative energy densities Ω_i as our dynamical variables, it is easy to interpret those fixed points. At R, the radiation dominates and normal matter is absent. Likewise, at M, the normal matter dominates while radiation is absent. The point O contains neither radiation nor matter, and is therefore dominated by the cosmological term because of (2.11c).

The Jacobian matrix of system (2.11) is computed straightforwardly. Evaluated at the three fixed points, we find

$$J(O) = \begin{pmatrix} -3 & 0 \\ 0 & -4 \end{pmatrix}, \quad J(R) = \begin{pmatrix} 1 & 0 \\ 3 & 4 \end{pmatrix}, \quad J(M) = \begin{pmatrix} 3 & 4 \\ 0 & -1 \end{pmatrix},$$

(2.12)

respectively. The corresponding eigenvalues of the stability matrix are given by

$$O: \quad \lambda_1 = -3, \quad \lambda_2 = -4, \tag{2.13a}$$

$$R: \quad \lambda_1 = 1, \quad \lambda_2 = 4, \tag{2.13b}$$

$$M: \quad \lambda_1 = -1, \quad \lambda_2 = 3, \tag{2.13c}$$

which implies that O is the only attractor of the system. Therefore, all trajectories will eventually approach O. R is unstable, however, since both eigenvalues are positive, we can think of R as the only past-time attractor. This means all trajectories will have 'started' at R. Lastly, M is a saddle point. This means that some trajectories are attracted towards M but are eventually repelled to move towards O. The phase space diagram Fig. 2 clearly shows these features.

In the cosmological context this has the following interpretation. Consider a spatially flat universe filled with normal matter and radiation, and with a very small cosmological term.[c] Such a universe will generically be dominated by radiation at early times, then it

---

[c]If the cosmological term happens to be 'large' then matter will never dominate and one obtains an almost direct transition from radiation to a state where the cosmological term dominates.

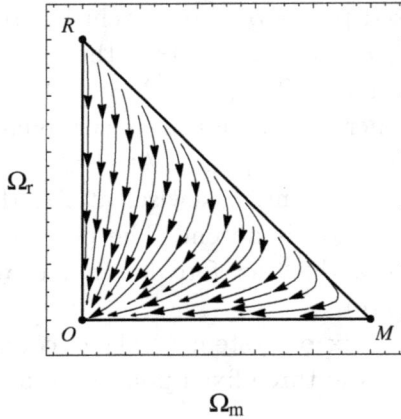

Fig. 2.   Phase space diagram of system (2.11).

will undergo a period where matter dominates its energy contents. Eventually, it will evolve to a state where the cosmological term dominates. This result is in line with our expectation of a good cosmological model, see (1.11).

2.2.   *Lyapunov functions*

The following method of studying the stability of a fixed point goes back to Lyapunov. It is completely different to linear stability and can be applied directly to the system in question. The main problem with this approach is that one has to be able to guess the Lyapunov function since there is no systematic way of doing so. Let us start by defining what a Lyapunov function is and its relation to stability of an autonomous system of equations.

Definition 2.5 (Lyapunov function). Let $\dot{\mathbf{x}} = \mathbf{f}(\mathbf{x})$ with $\mathbf{x} \in X \subset \mathbb{R}^n$ be a smooth autonomous system of equations with fixed point $\mathbf{x}_0$. Let $V : \mathbb{R}^n \to \mathbb{R}$ be a continuous function in a neighbourhood U of $\mathbf{x}_0$, then V is called a Lyapunov function for the point $\mathbf{x}_0$ if

(1) V is differentiable in $U \backslash \{\mathbf{x}_0\}$,
(2) $V(\mathbf{x}) > V(\mathbf{x}_0)$,
(3) $\dot{V} \leq 0, \forall x \in U \backslash \{\mathbf{x}_0\}$.

Note that the third requirement is the crucial one. It implies

$$\frac{d}{dt}V(x_1, x_2, \ldots, x_n) = \frac{\partial V}{\partial x_1}\dot{x}_1 + \cdots + \frac{\partial V}{\partial x_n}\dot{x}_n$$
$$= \frac{\partial V}{\partial x_1}f_1 + \cdots + \frac{\partial V}{\partial x_n}f_n \leq 0, \qquad (2.14)$$

which required repeated use of the chain rule and substitution of the autonomous system equations to eliminate the terms $\dot{x}_i$ for $i = 1, \ldots, n$.

One can conveniently write dV/dt using vector calculus notation

$$\frac{d}{dt}V(x_1, x_2, \ldots, x_n) = \text{grad}\, V \cdot \dot{\mathbf{x}} = \text{grad}\, V \cdot \mathbf{f}(\mathbf{x}). \qquad (2.15)$$

Let us now state the main theorem which connects a Lyapunov function to the stability of a fixed point of a dynamical system.

Theorem 2.6 (Lyapunov stability). *Let $\mathbf{x}_0$ be a critical point of the system $\dot{\mathbf{x}} = \mathbf{f}(\mathbf{x})$, and let U be a domain containing $\mathbf{x}_0$. If there exists a Lyapunov function $V(\mathbf{x})$ for which $\dot{V} \leq 0$, then $\mathbf{x}_0$ is a stable fixed point. If there exists a Lyapunov function $V(\mathbf{x})$ for which $\dot{V} < 0$, then $\mathbf{x}_0$ is a asymptotically stable fixed point.*

Furthermore, if $\|\mathbf{x}\| \to \infty$ and $V(\mathbf{x}) \to \infty$ for all $\mathbf{x}$, then $\mathbf{x}_0$ is said to be globally stable or globally asymptotically stable, respectively.

One can also find some instability results, see e.g. Ref. 5, which will also depend on our ability to find a suitable Lyapunov function. However, we will not use results along those lines since we are mainly concerned about the stability of certain fixed points in the context of cosmology.

Should we be able to find a Lyapunov function satisfying the criteria of the Lyapunov stability theorem, we could establish (asymptotic) stability without any reference to a solution of the ODEs. However, just because we failed in finding a Lyapunov function at a particular point, this does not necessarily imply that such a point is unstable. Since there is no systematic way of constructing a function, it is possible that we were simply not clever enough to find a Lyapunov function for the critical point concerned.

A first example

This first example is taken from Ref. 4. Suppose that a system is described by the vector field

$$\dot{x} = y, \tag{2.16a}$$

$$\dot{y} = -x + \epsilon x^2 y, \tag{2.16b}$$

which has one critical point at $(x, y) = (0, 0)$. A candidate Lyapunov's function is given by

$$V(x, y) = \frac{x^2 + y^2}{2}, \tag{2.17}$$

satisfying $V(0, 0) = 0$ and $V(x, y) > 0$ in the neighbourhood of the fixed point. This function leads to

$$\dot{V} = \operatorname{grad} V \cdot (\dot{x}, \dot{y}) = \epsilon x^2 y^2 \tag{2.18}$$

from which we conclude that the point is globally asymptotically stable if $\epsilon < 0$ since $x^2 y^2$ is positive definite and thus $\dot{V} < 0$ in the neighbourhood of the fixed point. It is important to emphasise, however, that $\epsilon > 0$ does not imply instability.

A second example

Let us consider the system

$$\dot{x} = -x^3 + xy, \tag{2.19a}$$

$$\dot{y} = -y - 2x^2 - x^2 y, \tag{2.19b}$$

which has one fixed point at $(x, y) = (0, 0)$. Computing the eigenvalues of the Jacobian matrix at the fixed point yields $\lambda_1 = -1$ and $\lambda_2 = 0$. Therefore, we cannot decide, based on linear stability theory, whether the origin is stable or not. However, starting with the candidate Lyapunov function

$$V(x, y) = 2x^2 + y^2 \tag{2.20}$$

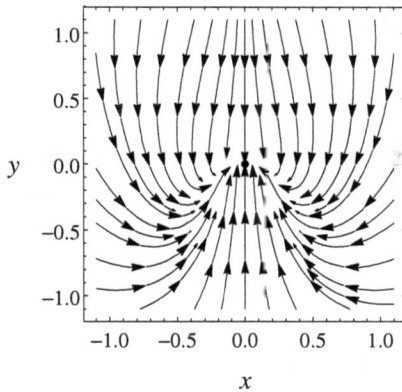

Fig. 3.   Phase space plot of the system (2.19).

leads to

$$\dot{V} = -4x^4 - 2y^2 - 2x^2y^2. \tag{2.21}$$

Therefore the point is globally asymptotically stable since all terms in $\dot{V}$ are negative definite and thus $\dot{V} < 0$ in the neighbourhood of the fixed point. This example has been adapted from a similar one in Ref. 6.

Note that the phase space plot is in agreement with our conclusion of stability, see Fig. 3.

Exercise 1.3. Consider the damped harmonic oscillator $\ddot{q} + 2\zeta\dot{q} + \omega^2 q = 0$. Rewrite this as a dynamical system, find its critical points and their stability. Find a Lyapunov function. Can you interpret the Lyapunov function.

2.3.   *Centre manifold theory*

Centre manifold theory is a method that allows us to simplify dynamical systems by reducing their dimensionality near fixed points with vanishing eigenvalues of the Jacobian matrix. It is also central to other elegant concepts such as bifurcations and the method of normal forms.[7] Here the essential basics of centre manifold theory are discussed following Refs. 4 and 6.

Let us, as above, consider the dynamical system

$$\dot{\mathbf{x}} = \mathbf{f}(\mathbf{x}) \qquad (2.22)$$

with $\mathbf{x} \in \mathbb{R}^n$ and let us assume that it has a fixed point $\mathbf{x}_0$. Near this point we can linearise the system using (2.7). Denoting $\mathbf{y} = \mathbf{x} - \mathbf{x}_0$, we can write (2.22)

$$\dot{\mathbf{y}} = J\mathbf{y}, \qquad (2.23)$$

where we emphasise that J is a constant coefficient $n \times n$ matrix. As such it will have n eigenvalues which motivate the following. The space $\mathbb{R}^n$ is the direct sum of three subspaces which are denoted by $\mathbb{E}^s$, $\mathbb{E}^u$ and $\mathbb{E}^c$, where the superscripts stand for stable, unstable and centre, respectively. The space $\mathbb{E}^s$ is spanned by the eigenvectors of J which have negative real part, $\mathbb{E}^u$ is spanned by the eigenvectors of J which have positive real part, and $\mathbb{E}^c$ is spanned by the eigenvectors of J which have zero real part. Linear stability theory is sufficient to understand the dynamics of trajectories in $\mathbb{E}^s$ and $\mathbb{E}^u$. Centre manifold theory will determine the dynamics of trajectories in $\mathbb{E}^c$.

In the context of centre manifold theory, it is useful to write our dynamical system (2.22) in the form

$$\dot{\mathbf{x}} = A\mathbf{x} + f(\mathbf{x}, \mathbf{y}), \qquad (2.24a)$$
$$\dot{\mathbf{y}} = B\mathbf{y} + g(\mathbf{x}, \mathbf{y}), \qquad (2.24b)$$

where $(x, y) \in \mathbb{R}^c \times \mathbb{R}^s$. Moreover, we assume

$$f(0,0) = 0, \qquad \nabla f(0,0) = 0, \qquad (2.25a)$$
$$g(0,0) = 0, \qquad \nabla g(0,0) = 0. \qquad (2.25b)$$

In system (2.24), A is a $c \times c$ matrix having eigenvalues with zero real parts, while B is an $s \times s$ matrix whose eigenvalues have negative real parts. Our aim is to understand the centre manifold of this system in order to investigate its dynamics. We have suppressed some regularity assumptions on f and g for simplicity.

Definition 2.7 (Centre Manifold). A geometrical space is a centre manifold for (2.24) if it can be locally represented as

$$W^c(0) = \{(x,y) \in \mathbb{R}^c \times \mathbb{R}^s | y = h(x), |x| < \delta, h(0) = 0, \nabla h(0) = 0\} \tag{2.26}$$

for δ sufficiently small.

The conditions $h(0) = 0$ and $\nabla h(0) = 0$ from the definition imply that the space $W^c(0)$ is tangent to the eigenspace E^c at the critical point $(x, y) = (0, 0)$.

Centre manifold theory is based on three main theorems.[4] The first one is about the existence of the centre manifold, the second one clarifies the issue of stability of solutions while the last one is about constructing the actual centre manifold needed to investigate the stability. We will state those theorems but will not state the proofs, the interested reader is referred to Ref. 6.

Theorem 2.8 (Existence). *There exists a centre manifold for (2.24). The dynamics of the system (2.24) restricted to the centre manifold is given by*

$$\dot{u} = Au + f(u, h(u)) \tag{2.27}$$

for $u \in \mathbb{R}^c$ sufficiently small.

Theorem 2.9 (Stability). *Suppose the zero solution of (2.27) is stable (asymptotically stable or unstable). Then the zero solution of (2.24) is also stable (asymptotically stable or unstable). Furthermore, if $(x(t), y(t))$ is also a solution of (2.24) with $(x(0), y(0))$ sufficiently small, there exists a solution $u(t)$ of (2.27) such that*

$$x(t) = u(t) + \mathcal{O}(e^{-\gamma t}), \tag{2.28a}$$
$$y(t) = h(u(t)) + \mathcal{O}(e^{-\gamma t}) \tag{2.28b}$$

as $t \to \infty$, where $\gamma > 0$ is a constant.

We now know that the centre manifold exists, and we can establish the stability or instability of a solution. However, our ability to do so depends on the knowledge of the function $h(x)$ in Definition 2.7. We will now derive a differential equation for the function $h(x)$.

Following Definition 2.7, we have that $y = h(x)$. Let us differentiate this with respect to time and apply the chain rule. This gives

$$\dot{y} = \nabla h(x) \cdot \dot{x}. \tag{2.29}$$

Since $W^c(0)$ is based on the dynamics generated by the system (2.24), we can substitute for $\dot{x}$ the right-hand side of (2.24a) and for $\dot{y}$ the right-hand side of (2.24b). This yields

$$Bh(x) + g(x, h(x)) = \nabla h(x) \cdot [Ax + f(x, h(x))], \tag{2.30}$$

where we also used that $y = h(x)$. The latter equation can be re-arranged into the quasilinear partial different equation

$$\mathcal{N}(h(x)) := \nabla h(x)[Ax + f(x, h(x))] - Bh(x) - g(x, h(x)) = 0 \tag{2.31}$$

which must be satisfied by $h(x)$ for it to be the centre manifold. In general, we cannot find a solution to this equation. Even for relatively simple dynamical systems it is often impossible to find an exact solution of this equation. It is the third and last theorem which explains why not all is lost at this point.

Theorem 2.10 (Approximation). *Let $\phi : \mathbb{R}^c \to \mathbb{R}^s$ be a mapping with $\phi(0) = \nabla\phi(0) = 0$ such that $\mathcal{N}(\phi(x)) = \mathcal{O}(|x|^q)$ as $x \to 0$ for some $q > 1$. Then*

$$|h(x) - \phi(x)| = \mathcal{O}(|x|^q) \quad as \ x \to 0. \tag{2.32}$$

The main point of this theorem is that an approximate knowledge of the centre manifold returns the same information about stability as the exact solution of Eq. (2.31). It turns out that finding an approximation for the centre manifold is a fairly doable task in comparison to finding the exact solution. The centre manifold machinery is best explained with a concrete example.

Example — a simple two-dimensional model

The following two-dimensional example is taken from Ref. 4. We consider the system

$$\dot{x} = x^2 y - x^5, \tag{2.33a}$$
$$\dot{y} = -y + x^2. \tag{2.33b}$$

The origin $(x, y) = (0, 0)$ is a fixed point. The Jacobian matrix of the linearised system about the origin has eigenvalues of 0 and -1. Since there is a zero eigenvalue, the point is non-hyperbolic and linear stability theory fails to determine the nature of stability of this point.

By Theorem 2.8, there exists a centre manifold for the system (2.33) and it can be represented locally as

$$W^c(0) = \{(x, y) \in \mathbb{R}^2 | y = h(x), |x| < \delta, h(0) = Dh(0) = 0\} \tag{2.34}$$

for δ sufficiently small. Next, we need to compute $W^c(0)$. Here we can exploit Theorem 2.10 which says that it suffices to approximate the centre manifold to establish stability properties. Therefore, it is customary to assume an expansion for $h(x)$ of the form

$$h(x) = ax^2 + bx^3 + \mathcal{O}(x^4), \tag{2.35}$$

where a and b are constants to be determined. This expression is then substituted into (2.31) with the aim of determining those constants.

In this example, Eqs. (2.33) yield

$$A = 0, \quad B = -1, \tag{2.36a}$$
$$f(x, y) = x^2 y - x^5, \tag{2.36b}$$
$$g(x, y) = x^2. \tag{2.36c}$$

This, in addition to (2.35), is substituted into (2.31) and gives

$$\mathcal{N} = (2ax + 3bx^2 + \cdots)(ax^4 + bx^5 - x^5 + \cdots)$$
$$+ ax^2 + bx^3 - x^2 + \cdots = 0. \tag{2.37}$$

The coefficients of each power of x must be zero so that (2.37) holds. This provides us with a set on linear equations in the constants a

and b which are solved by

$$a = 1, \quad b = 0, \tag{2.38}$$

where all terms of order $\mathcal{O}(x^4)$ have been ignored. Therefore, the centre manifold is locally given by

$$h(x) = x^2 + \mathcal{O}(x^4). \tag{2.39}$$

Finally, following Theorem 2.8, the dynamics of the system restricted to the centre manifold is obtained to be

$$\dot{x} = x^4 + \mathcal{O}(x^5). \tag{2.40}$$

We conclude that for x sufficiently small, $x = 0$ is unstable. Therefore, the critical point $(0, 0)$ is unstable. In Fig. 4, we show the phase space for this system and also indicate the centre manifold.

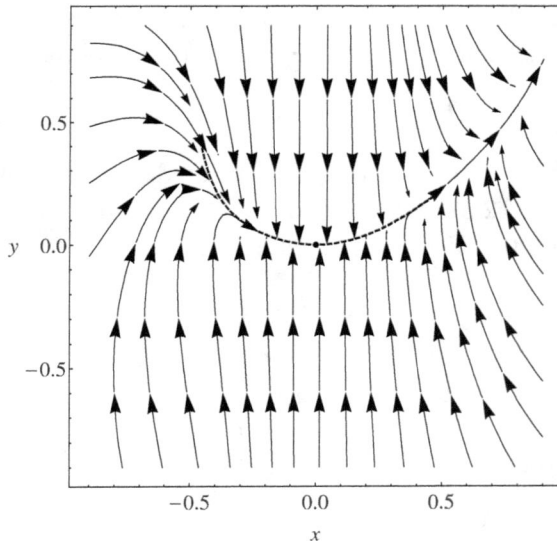

Fig. 4. Phase space plot of the system (2.33). The centre manifold is indicated by a dashed line and was computed up to terms x^{13}. One sees very clearly how the centre manifold attracts the trajectories and how they are repelled from the origin (along the centre manifold) making this point unstable.

Exercise 1.4. Find the critical points of $x' = x^2 y + 2x^2 - x^5$, $y' = -y + 2 + x^2$ and study the stability of the non-hyperbolic critical point.

3. Cosmology Using Dynamical Systems

We discussed some aspects of cosmology in the context of dynamical systems, see Refs. 2 and 3 for more details, and also Refs. 8 or 9 for anisotropic models. Based on the series of cosmological epochs inflation $\rightarrow$ radiation $\rightarrow$ matter $\rightarrow$ cosmological term of Section 1.3, which could be called a 'minimal' cosmological model, we will now make links with dynamical systems. A very neat paper studying cosmological models in the Lotka–Volterra framework is Ref. 10.

Let us now consider a generic 'minimal' cosmological model described by an $n \times n$ system of autonomous equations. Should this model begin with an inflationary period, then this should correspond to an early-time attractor in the dynamical system. All eigenvalues of the Jacobian matrix at this point should be positive in order to ensure that all trajectories evolve away from this point, this means $\lambda_i > 0$ for $i = 1, \ldots, n$.

In an ideal model we would also have two saddle points ($\lambda_j > 0$, $\lambda_k < 0$ with $j + k = n$) which correspond to a radiation dominated and matter dominated universe, respectively. These epochs being saddle points makes sure that some trajectories are attracted to these points, however, they will eventually be repelled. In this case the universe will evolve through both epochs. Let us note here that most models will only contain either matter or radiation, and thus we would be satisfied if there was only one saddle point.

Lastly, we require a late-time attractor ($\lambda_i < 0$ for $i = 1, \ldots, n$) where the universe is undergoing an accelerated expansion which corresponds to the de Sitter solution. We say the universe is approaching de Sitter space asymptotically. This can be summarised as follows

$$\text{inflation} \longrightarrow \text{radiation/matter} \longrightarrow \text{de Sitter}$$
$$\lambda_i > 0 \qquad \lambda_j > 0, \lambda_k < 0 \qquad \lambda_i < 0 \tag{3.1}$$

where, for simplicity, we neglected the possibility of some zero eigen-values.

3.1.   *Cosmology with matter and scalar field*

The cosmological constant Λ has strong observational support, but also leads to a variety of problems which are called the cosmological constant problems, we refer the reader in particular to Ref. 11. These problems can largely be avoided if the constant term Λ is replaced by a dynamically evolving scalar field φ with some given potential $V(\varphi)$. In this case one often speaks of dark energy. In many models the potential V is assumed to be of exponential form, $V = V_0 \exp(-\lambda \kappa \varphi)$.

Moreover, instead of writing the equation of state for the matter as $p = w\rho$, one often encounters a slightly different parametrisation which is given by

$$p_\gamma = w_\gamma \rho_\gamma = (\gamma - 1)\rho_\gamma, \tag{3.2}$$

where $\gamma = 1 + w_\gamma$ is a constant and $0 \le \gamma \le 2$. Its value is $4/3$ when there is radiation, and is 1 for standard matter or dark matter in this context.

For this setup, the Einstein field equations are

$$H^2 = \frac{\kappa^2}{3}\left(\rho_\gamma + \frac{1}{2}\dot{\varphi}^2 + V\right), \tag{3.3a}$$

$$\dot{H} = -\frac{\kappa^2}{2}(\rho_\gamma + p_\gamma + \dot{\varphi}^2). \tag{3.3b}$$

We can interpret $\rho_\varphi = \dot{\varphi}^2/2 + V$ as the energy density of the scalar field and $p_\varphi = \dot{\varphi}^2/2 - V$ as its pressure. This also allows us to define an effective equation of state for the field. The conservation equations for the matter and the scalar field are given by

$$\dot{\rho}_\gamma = -3H(\rho_\gamma + p_\gamma), \tag{3.4a}$$

$$\ddot{\varphi} = -3H\dot{\varphi} - \frac{dV}{d\varphi} = -3H\dot{\varphi} + \lambda\kappa V, \tag{3.4b}$$

where we used the exponential form of the potential. We follow the approach outlined in Section 1.4 and rewrite (3.3) and (3.4) using more suitable variables. As before, we start with dividing Eq. (3.3a) with H^2 which results in

$$1 = \frac{\kappa^2 \rho_\gamma}{3H^2} + \frac{\kappa^2 \dot{\varphi}^2}{6H^2} + \frac{\kappa^2 V}{3H^2}. \tag{3.5}$$

Every term on the right-hand side is positive since $V > 0$ and $\rho_\gamma > 0$, and it turns out that the following the dimensionless variables[12,13] are particularly useful

$$x^2 = \frac{\kappa^2 \dot{\varphi}^2}{6H^2}, \quad y^2 = \frac{\kappa^2 V}{3H^2}, \quad s^2 = \frac{\kappa^2 \rho_\gamma}{3H^2}, \tag{3.6}$$

which transform (3.5) into

$$1 = x^2 + y^2 - s^2. \tag{3.7}$$

Therefore, we can choose x, y as two independent variables. This leads to

$$1 \geq 1 - x^2 - y^2 = s^2 = \frac{\kappa^2 \rho_\gamma}{3H^2} \geq 0 \tag{3.8}$$

implying that $0 \leq x^2 + y^2 \leq 1$ which means that the physical phase space of this model is contained within the unit circle.

We will introduce three more quantities which are useful in understanding the physical properties at the fixed points. The dimensionless density parameter (1.7) of the scalar field φ can be expressed in terms of the new variables and is given by

$$\Omega_\varphi = \frac{\kappa^2 \rho_\varphi}{3H^2} = x^2 + y^2. \tag{3.9}$$

Moreover, we define the equation of state for the scalar field by

$$\gamma_\varphi = 1 + w_\varphi = 1 + \frac{p_\varphi}{\rho_\varphi} = \frac{2x^2}{x^2 + y^2}. \tag{3.10}$$

Lastly, we define the effective equation of state of the total system by

$$\begin{aligned} w_{\text{eff}} &= \frac{p_\gamma + p_\varphi}{\rho_\gamma + \rho_\varphi} = \frac{w_\gamma \rho_\gamma + \dot{\varphi}^2/2 - V}{\rho_\gamma + \dot{\varphi}^2/2 + V} \\ &= w_\gamma (1 - x^2 - y^2) + x^2 - y^2. \end{aligned} \tag{3.11}$$

Now, we are ready to derive a two-dimensional dynamical system using the variables x and y. As before, we will introduce a new 'time' variable $N = \log(a)$ so that $dN = Hdt$, and denote differentiation with respect to N by a prime.

Let us begin by differentiating x with respect to time t

$$\dot{x} = \frac{\kappa}{\sqrt{6}} \frac{\ddot{\varphi}H - \dot{\varphi}\dot{H}}{H^2} = \frac{\kappa}{\sqrt{6}} \left(\frac{\ddot{\varphi}}{H} - \dot{\varphi}\frac{\dot{H}}{H^2} \right). \tag{3.12}$$

Substituting for $\ddot{\varphi}$ using (3.4b) and for $\dot{H}$ using (3.3b) we arrive at

$$\dot{x} = \frac{\kappa}{\sqrt{6}} \left(-3\dot{\varphi} + \lambda\kappa\frac{V}{H} + \dot{\varphi}\frac{\kappa^2}{2H^2}(\gamma\rho_\gamma + \dot{\varphi}^2) \right). \tag{3.13}$$

Next, using the variables (3.6) and the condition (3.7) we get

$$\dot{x} = H \left[-3x + \sqrt{\frac{3}{2}}\lambda y^2 + \frac{3}{2}x \left((1 - x^2 - y^2)\gamma + 2x^2 \right) \right]. \tag{3.14}$$

One can now introduce the new 'time parameter' N. Following similar steps, the equation for y' can be derived. The final system is

$$x' = -3x + \sqrt{\frac{3}{2}}\lambda y^2 + \frac{3}{2}x(2x^2 + \gamma(1 - x^2 - y^2)), \tag{3.15a}$$

$$y' = -\lambda\sqrt{\frac{3}{2}}xy + \frac{3}{2}y \left(2x^2 + \gamma(1 - x^2 - y^2) \right). \tag{3.15b}$$

The complete dynamics of this cosmological model is describe by (3.15).

We noted that the phase space of this system is contained in the unit circle. Inspection of the dynamical equations shows that system (3.15) is invariant under the transformation $y \mapsto -y$ and symmetric under time reversal $t \mapsto -t$. This implies that we can restrict our analysis on the upper half-disc with $y > 0$. The lower half-disc of the phase space corresponds to the contracting universe because $H < 0$ in this region.

The properties of the dynamical system (3.15) depend on the values of the constants λ and γ. Amongst others, they will in particular affect the existence and stability of the fixed points of the system,

Table 2. Critical point of the system (3.15).

	x	y	Existence
O	0	0	$\forall \lambda$ and γ
A_+	1	0	$\forall \lambda$ and γ
A_-	-1	0	$\forall \lambda$ and γ
B	$\lambda/\sqrt{6}$	$[1 - \lambda^2/\epsilon]^{1/2}$	$\lambda^2 < 6$
C	$\sqrt{3/2}\gamma/\lambda$	$[3(2-\gamma)\gamma/2\lambda^2]^{1/2}$	$\lambda^2 > 3\gamma$

Table 3. Summary of the properties of the critical points.

	Stability	Ω_φ	γ_φ
O	saddle point for $0 < \gamma < 2$	0	Undefined
A_+	unstable node for $\lambda < \sqrt{6}$ and saddle point for $\lambda > \sqrt{6}$	1	2
A_-	unstable node for $\lambda > -\sqrt{6}$ and addle point for $\lambda < -\sqrt{6}$	1	2
B	stable node for $\lambda^2 < 3\gamma$ and saddle pcint for $3\gamma < \lambda^2 < 6$	1	$\lambda^2/3$
C	stable node for $3\gamma < \lambda^2 < 24\gamma^2/(9\gamma - 2)$ and stable spiral for $\lambda^2 > 24\gamma^2/(9\gamma - 2)$	$3\gamma/\lambda^2$	γ

see Ref. 12. This can be related to the theory of bifurcations, something that has not been explored in cosmological dynamical systems. Table 2 contains all critical points of the system (3.15).

Having found all the possible fixed points, we can now compute the eigenvalues and determine their stability which is summarised in Table 3, see Ref. 12.

Figure 5 shows the phase spaces of this model for a particular parameter choice.

It should be noted that the inequality signs in Table 3 exclude certain values from the analysis. For instance, when we choose $\lambda^2 = 3\gamma$, the two points B and C have the same coordinates (the system has one critical point less), namely $x_0 = \sqrt{\gamma/2}$ and $y_0 = \sqrt{1 - \gamma/2}$ so that $x_0^2 + y_0^2 = 1$ and its eigenvalues are $0, 3/2(\gamma - 2)$. Linear stability theory cannot determine the stability of this point. One could, in

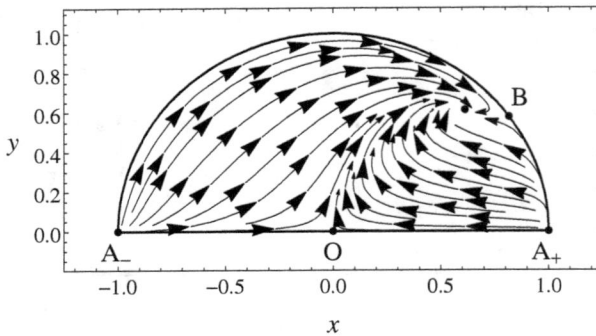

Fig. 5.   Phase space plot scalar field cosmology with exponential potential and matter. Parameter values are $\gamma = 1$ and $\lambda = 2$.

principle, apply centre manifold theory. However, this is problematic as the physical phase space is bounded by the unit circle and centre manifold theory will take into account the entire phase space. One could construct the centre manifold and only consider it inside the circle but this also has problems. For concreteness we set $\gamma = 1$ in the following, which means $\lambda = \sqrt{3}$ and $x_0 = y_0 = \sqrt{1/2}$.

The easiest way forward is to use Lyapunov's method near this point. We start with the candidate Lyapunov function of the form

$$V = \left(x - \frac{1}{\sqrt{2}}\right)^2 + 4\left(y - \frac{1}{\sqrt{2}}\right)^2 \qquad (3.16)$$

and one verify that this function satisfies $\dot{V} < 0$ near the critical point. Since the function is positive definite near that point by construction, we can apply Theorem 2.6. Following for instance Ref. 5, we can estimate the region of asymptotic stability. Defining $S_\delta := \{(x, y)|V \leq \delta\}$ for $\delta \geq 0$, and denoting by C_δ the component of S_δ containing the critical point, we have the following statement.[5] Let Ω be the set where $\dot{V} < 0$, then the interior of C_δ contained in Ω lies in the region of asymptotic stability. As mentioned earlier, this approach relies on our ability to find a suitable Lyapunov function. Different choices can result in different parts of the region of

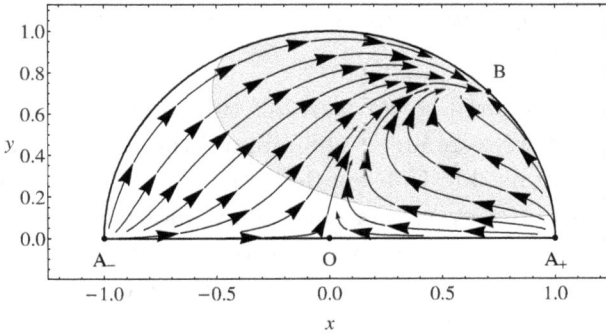

Fig. 6. Phase space plot scalar field cosmology with exponential potential and matter. Parameter values are $\gamma = 1$ and $\lambda = \sqrt{3}$. The shaded area shows part of the region of asymptotic stability of the fixed point. In this region $\dot{V} < 0$ and $V < 3/2$.

asymptotic stability being covered and there is no guarantee that the entire region can be identified by this method alone. In Fig. 6, we show the region of asymptotic stability based on the Lyapunov function (3.16) for model (3.15). A better Lyapunov function would of course improve this picture and increase the region.

A detailed and comprehensive phase-space analysis, based on linear stability theory alone, of this model can be found in Ref. 12. Other methods were explored in Ref. 14. A complete discussion of all its properties in the context of cosmology is also given. This model has many interesting features as well as some problems which motivates various extensions, many of which have been considered in the literature. In fact, the literature of dynamical systems applications in early-time and late-time cosmology is so vast, that it could fill several books with ease!

We should point out that this model falls short our wish list (3.1). The early-time fixed points $A_\pm$ are dominated by the scalar field, however, the effective equation of state is $w_{\text{eff}} = 1$ which is unphysical. It is point C which makes this model so interesting because this fixed point is stable and contains both, a non-vanishing scalar field and matter. One speaks of scaling solutions as the scalar field energy density is proportional to that of the fluid.

3.2. *Cosmology with matter and scalar field and interactions*

The models considered so far were all two dimensional. This relied on the fact that we were able to 'eliminate' the Hubble parameter H from the equations due to a smart choice of variables and a clever choice of 'time'. However, there are many known models where this approach does not work and one has to introduce new variables. In the following we will discuss one such type of models and a possible choice of a new variable.

The cosmological Einstein field equations (3.3)–(3.4) are compatible with the introduction of an additional interaction term Q, say. This interaction would allow for an energy transfer from the scalar φ to the matter ρ_γ and vice versa. The introduction of such a term leaves Eqs. (3.3) unchanged, but (3.4) becomes

$$\dot{\rho}_\gamma = -3H(\rho_\gamma + p_\gamma) - Q, \tag{3.17a}$$

$$\ddot{\varphi} = -3H\dot{\varphi} - \frac{dV}{d\varphi} + \frac{Q}{\dot{\varphi}}, \tag{3.17b}$$

where we note that the term $Q\dot{\varphi}$ is natural when one computes the conservation equation $\dot{\rho}_\varphi = -3H(\rho_\varphi + p_\varphi)$. Various choices for the coupling function Q were considered in the literature, for instance $Q = \alpha H \rho_\gamma$ or $Q = (2/3)\kappa\beta\rho_\gamma\dot{\varphi}$ with α and β being dimensionless constants whose sign determines the direction of energy transfer from one component to the other.[15–17] Those two choices can be motivated physically, however, one of the main motivation is the fact that the dynamical system with these coupling remains two dimensional as the Hubble parameter can be eliminated from the equations. However, both choices appear rather arbitrary and one would prefer a choice where the coupling is simply proportional to an energy density, for instance $Q = \Gamma\rho_\gamma$ with Γ assumed to be small, see Ref. 18, or for a further generalisation Ref. 19. In this case the phase space cannot be represented in the plane and one has to work in a three-dimensional space.

As before, we start with the variables (3.6) but need a third variable in order to be able to write the cosmological field equations as an autonomous system of differential equations. A possible third variable z can be chosen to be

$$z = \frac{H_0}{H + H_0}, \tag{3.18}$$

where H_0 is the Hubble parameter at an arbitrary fixed time. It is convenient to choose this time to be 'today'. This variable z ensures that the physical phase-space is compact. The Hubble parameter $H \to 0$ in the early-time universe and $H \to \infty$ for the late-time universe. Therefore

$$z = \begin{cases} 0 & \text{if } H = 0, \\ 1/2 & \text{if } H = H_0, \\ 1 & \text{if } H \to \infty, \end{cases} \tag{3.19}$$

and z is bounded by $0 \le z \le 1$. Since the phase-space of system (3.15) is half a unit circle, we have that with coupling term $Q = \Gamma \rho_\gamma$ the phase-space now corresponds to a half-cylinder of unit height and unit radius.

The resulting dynamical system is given by

$$x' = -3x + \lambda \frac{\sqrt{6}}{2} y^2 + \frac{3}{2}x(1 + x^2 - y^2) - \zeta \frac{(1 - x^2 - y^2)z}{2x(z-1)}, \tag{3.20a}$$

$$y' = -\lambda \frac{\sqrt{6}}{2} xy + \frac{3}{2}y(1 + x^2 - y^2), \tag{3.20b}$$

$$z' = \frac{3}{2}z(1 - z)(1 + x^2 - y^2), \tag{3.20c}$$

where $\zeta = \Gamma/H_0$. A detailed phase-space analysis of this model can be found in Ref. 18.

Exercise 1.5. Derive the results summarised in Tables 2 and 3 and show that there are no additional critical points.

Exercise 1.6. Study the system (3.20) and carefully consider the critical points in the limit $z \to 1$.

4. Final Remarks

It is hoped that this chapter succeeded in giving the reader a useful introduction into the exciting field of dynamical systems in cosmology. We should remark that the majority of papers dealing with the subject are confined to linear stability theory and focus more on the interpretation of results in the context of cosmology. However, there are many models where a more in-depth analysis is needed to gain a complete understanding of the physics involved. Moreover, there is no need to select models primarily because of their simpler mathematical structure since we have all the tools at hand to study the more difficult ones too. We hope the reader feels encouraged to study all aspects of a cosmological dynamical system and use a variety of techniques developed by mathematicians, beyond linear stability theory. As Einstein wrote 'Everything should be made as simple as possible, but not simpler.'

5.   Further Reading

For a general introduction to cosmology Ref. 1 is an accessible book which includes various topics important for modern cosmology. The books, Refs. 2 and 3, provide readers with good introductions to the use of dynamical systems in cosmology, while Ref. 4 deals extensively with dynamical systems from a more applied mathematics point of view. Of particular interest to the reader might be Ref. 10 which studies cosmological dynamical systems from the point of view of the Lotka–Volterra framework. New articles appear in this field on a regular basis because of the power of this approach. A comprehensive dynamical systems analysis of a given cosmological model can often determine whether such a model should be investigated further or is already excluded from being to able to describe the universe.

6.   Solutions to Selected Exercises

Exercise 1.1 Solution: For $w = 0$ the conservation equation (1.3) is solved by $\rho = \rho_0 a^{-3}$ where ρ_0 is a constant of integration. Then we can write Eq. (1.2a) in the form $\dot{a} = \sqrt{\kappa \rho_0/(3a) - 1}$ which we can solve using separation of variables and integration. One cannot find an explicit solutions $a(t)$ but can parametrise the solution as follows: $a = \kappa\rho_0/6(1 - \cos(u))$ and $t = \kappa\rho_0/6(u - \sin(u))$ which is the standard parametrisation of the cycloid.

Exercise 1.3 Solution: We introduce $\dot{q} = p$ so that $\dot{p} = -\omega^2 q - 2\zeta p$. The critical point is $q = p = 0$, with eigenvalues $-\zeta \pm \sqrt{\zeta^2 - \omega^2}$. Consider $V = \omega q^2/2 + p^2/2$ which gives $\dot{V} = -2\zeta p^2$ and this Lyapunov function is motivated by the energy of the harmonic oscillator. For $\zeta > 0$ this implies stability.

Exercise 1.4 Solution: Note that $y \mapsto y + 2$ relates this system to that discussed in the example preceding this exercise.

Acknowledgments

We would like to thank Nicola Tamanini and Matthew Wright for valuable comments on these notes.

References

1. S. Dodelson, *Modern Cosmology*. Academic Press, Elsevier, San Diego (2003).
2. J. Wainwright and G. F. R. Ellis, *Dynamical Systems in Cosmology*. Cambridge University Press (1997).
3. A. A. Coley, *Dynamical Systems and Cosmology*. Kluwer Academic Publishers, Dordrecht (2003).
4. S. Wiggins, *Introduction to Applied Nonlinear Dynamical Systems and Chaos*. Springer, New York (1990).
5. F. Brauer and J. A. Nohel, *The Qualitative Theory of Ordinary Differential Equations*. Dover Publications, New York (1989).
6. J. Carr, *Applications of Centre Manifold Theory*. Springer, New York Heidelberg Berlin (1981).
7. A. H. Nayfeh, *The Method of Normal Forms*. Wiley-VCH, Weinheim (2011).
8. A. D. Rendall, Cosmological models and centre manifold theory, *Gen. Rel. Grav.* **34**, 1277–1294 (2002).
9. A. Coley and S. Hervik, A Dynamical systems approach to the tilted Bianchi models of solvable type, *Class. Quant. Grav.* **22**, 579–606 (2005).
10. J. Perez, A. Füzfa, T. Carletti, L. Mélot and L. Guedezounme, The Jungle Universe: coupled cosmological models in a Lotka-Volterra framework, *Gen. Rel. Grav.* **46**, 1753 (2014).
11. J. Martin, Everything You Always Wanted To Know About The Cosmological Constant Problem (But Were Afraid To Ask), *Comptes. Rendus. Physique.* **13**, 566–665 (2012).

12. E. J. Copeland, A. R. Liddle and D. Wands, Exponential potentials and cosmological scaling solutions, *Phys. Rev. D* **57**, 4686–4690 (1998).

13. E. J. Copeland, M. Sami and S. Tsujikawa, Dynamics of dark energy, *Int. J. Mod. Phys. D* **15**, 1753–1936 (2006).

14. C. G. Böhmer, T. Harko and S. V. Sabau, Jacobi stability analysis of dynamical systems: Applications in gravitation and cosmology, *Adv. Theor. Math. Phys.* **16**, 1145–1196 (2012).

15. L. Amendola, Scaling solutions in general nonminimal coupling theories, *Phys. Rev. D* **60**, 043501 (1999).

16. D. J. Holden and D. Wands, Selfsimilar cosmological solutions with a nonminimally coupled scalar field, *Phys. Rev. D* **61**, 043506 (2000).

17. A. P. Billyard and A. A. Coley, Interactions in scalar field cosmology, *Phys. Rev. D* **61**, 083503 (2000).

18. C. G. Böhmer, G. Caldera-Cabral, R. Lazkoz and R. Maartens, Dynamics of dark energy with a coupling to dark matter, *Phys. Rev. D* **78**, 023505 (2008).

19. C. G. Böhmer, G. Caldera-Cabral, N. Chan, R. Lazkoz and R. Maartens, Quintessence with quadratic coupling to dark matter, *Phys. Rev. D* **81**, 083003 (2010).

Chapter 5

Lotka–Volterra Dynamical Systems

Stephen Baigent

Department of Mathematics, UCL, Gower Street,
London WC1E 6BT, UK
steve.baigent@ucl.ac.uk

Lotka–Volterra systems are used to introduce in a simple setting a number of dynamical systems techniques. Concepts such as omega limit sets, simple attractors, Lyapunov functions are explained in the context of Lotka–Volterra systems. We discuss LaSalle's Invariance Principle. Monotone systems theory is also introduced in the context of the Lotka–Volterra systems.

1. Introduction and Scope

The Lotka–Volterra equations are an important model that has been widely used by theoretical ecologists to study the implications of various interactions between members of a population in a fixed habitat containing a number of distinct interacting species. They are by no means the most realistic of such ecological models, but they are arguably the simplest since the highest order terms they involve are quadratic, and therefore they feature the next level of complexity up from linear. As we shall see, even amongst differential equations with quadratic terms, they have a very special form which makes them amenable to well-known mathematical techniques from standard linear algebra, convex analysis, and dynamical systems theory.

To set the scene, we write the Lotka–Volterra equations in the revealing form:

$$\frac{\dot{x}_i}{x_i} = r_i + \sum_{j=1}^{n} a_{ij} x_j, \tag{1.1}$$

where n is the number of distinct species, x_i is the population density of the ith species, and r_i, a_{ij} are all real numbers, possibly zero, and here assumed to be independent of time. Multiplying each equation through by x_i shows that indeed the equations are quadratic, but when written as above we see that the net population growth per individual per unit time $(\dot{x}_i/x_i)$ is linear in the population densities $x = (x_1, \ldots, x_n)$.

A good starting point in the study of the dynamics of (1.1) is to first locate steady states; that is, points x^* where $\dot{x}_i = 0$ for each $i = 1, \ldots, n$. Of special interest, since they model one scenario where all species can coexist, are the so-called *interior* steady states. These states satisfy $x_i^* > 0$ for each $i = 1, \ldots, n$ and so are obtained by solving the linear system

$$r_i + \sum_{j=1}^{n} a_{ij} x_j, \quad i = 1, \ldots, n. \tag{1.2}$$

Recall that the r_i, a_{ij} may be of any sign or zero. As we shall see, determining when (1.2) has a unique solution $x_i^* > 0$ for each $i = 1, \ldots, n$ relies heavily on linear algebraic techniques. All other steady states involve at least one density vanishing; that is at least one species is extinct. Such steady states are determined by investigating the linear system (1.2) with all possible subsets of $\{x_i\}_{i=1}^{n}$ set to zero.

The main virtue of model (1.1) is that it enables us to study on paper or on the computer the outcome of any set of interactions between the n species, and they are the simplest model that enables us to do so. The type of interactions we refer to are split into two categories: *intraspecific* (the effect of one member of a species on another member of the same species), and *interspecific* (the effect of a member of one species on a member of another species). The strength of the interactions are encoded in the parameters a_{ij}, which are usually assembled into the $n \times n$ interaction matrix $A = ((a_{ij}))$.

The parameters r_i determine the *intrinsic* growth rate per individual of species i which would be observed if intraspecific and interspecific interactions were absent. Here we will not be concerned with the precise details of the ecological or environmental mechanisms that contribute to the value of each of the parameters, as we are more interested in the effects of the signs and magnitudes of each parameter on the qualitative behaviour of (1.1).

We shall however, link the signs of parameters to types of interactions. For example, $r_i > 0$ says that the environment intrinsically favours the growth of species i, whereas $r_i < 0$ signals a risk of extinction for that species unless the presence of another species promotes its growth. An example of the latter case is where a predator will go extinct in the absence of its prey and a suitable substitute food source.

Much of the material will apply to the most general form of the Lotka–Volterra model (1.1). Existence of interior steady states will be investigated and their local stability studied.

2. Lyapunov Methods for Lotka–Volterra Systems

2.1. *Some basic dynamical systems results*

Some notation first: $\mathbb{R}_{\geq 0} = \{x \geq 0\}$, $\mathbb{R}_{>0} = \{x > 0\}$. We will always assume that parameters are such that the differential equations (1.1) generate a semiflow $\varphi_t : \mathbb{R}_{\geq 0}^n \to \mathbb{R}_{\geq 0}^n$: $\forall x \in \mathbb{R}_{\geq 0}^n$ and $s, t \geq 0$,

(1) $\varphi_0(x) = x$;
(2) $\varphi_t(\varphi_s(x)) = \varphi_{t+s}(x)$.

Let $U \subset \mathbb{R}^n$ be open.

Definition 2.1 (Orbit). The (forward) orbit of $x \in U$ is the set $O^+(x) = \{\varphi_t(x) : t \geq 0\}$.

Definition 2.2 (Steady state). A steady state of $\dot{x} = f(x)$ is a point $x \in U$ for which $f(x) = 0$.

Definition 2.3 (Forward invariant set). A set $S \subseteq U$ is a forward invariant set for φ_t if whenever $x \in S$ we have $\varphi_t(x) \in S$ for all $t \geq 0$.

Definition 2.4 (Invariant set). When φ_t is a flow (i.e. also defined for $t \leq 0$), the set $S \subseteq U$ is an invariant set for φ_t if whenever $x \in S$ we have $\varphi_t(x) \in S$ for all $t \in \mathbb{R}$.

One important use of invariant sets is captured by the following result.[2]

Theorem 2.5. *Let $S \subset \mathbb{R}^n$ be homeomorphic to the closed unit ball and forward invariant for the flow of $\dot{x} = f(x)$. Then the flow has a steady state $x^* \in S$.*

Hence one way of showing the existence of at least one steady state in a compact simply connected subset of $\mathbb{R}^n$ is to show that all orbits enter that set (so that it is forward invariant).

The Heine–Borel theorem states that a subset of $\mathbb{R}^n$ is compact if and only if it is closed and bounded. The key tool for studying the convergence of orbits is the *Omega limit set*. This is the totality of all limit points of the forward orbit through a given point. To prove that an orbit is convergent to a steady state, one needs to show that its omega limit set consists of a single point, namely that steady state. Other interesting limit sets are attracting limit cycles, periodic orbits, attractors, etc.

Definition 2.6 (Omega limit point). A point $p \in U$ is an omega limit point of $x \in U$ if there are points $\varphi_{t_1}(x), \varphi_{t_2}(x), \ldots$ on the orbit of x such that $t_k \to \infty$ and $\varphi_{t_k}(x) \to p$ as $k \to \infty$.

Definition 2.7 (Omega limit set). The omega limit set $\omega(x)$ of a point $x \in U$ under the flow φ_t is the set of all omega limit points of x.

There is a similar construct for when φ_t is defined backwards in time, such as when it is a flow.

Definition 2.8 (Alpha limit point). A point p is an α limit point for the point $x \in U$ if there are points $\varphi_{t_1}(x), \varphi_{t_2}(x), \ldots$ on the orbit of x such that $t_k \to -\infty$ and $\varphi_{t_k}(x) \to p$ as $k \to \infty$.

Definition 2.9 (Alpha limit set). The alpha limit set $\alpha(x)$ of a point $x \in U$ under the flow φ_t is the set of all alpha limit points of x.

Lemma 2.10 (Properties of Omega limit sets).

(1) $\omega(x)$ *is a closed set (but it might be empty).*
(2) *If* $\overline{O^+(x)}$ *is compact, then* $\omega(x)$ *is non-empty and connected.*
(3) $\omega(x)$ *is an invariant set for* φ_t.
(4) *If* $y \in O^+(x)$ *then* $\omega(y) = \omega(x)$.
(5) $\omega(x)$ *can be written as*

$$\omega(x) = \bigcap_{t\geq 0} \overline{\{\varphi_s(x) : s \geq t\}} = \bigcap_{t\geq 0} \overline{O^+(\varphi_t(x))},$$

where $\overline{A}$ *is the closure of* A.

For a proof, see, for example, Ref. 4.

Example 2.11. $\dot{x} = 1$ has the flow $\varphi_t(x) = x + t$. Given any $x \in \mathbb{R}$ and any sequence $t_k \to \infty$, $\varphi_{t_k}(x) \to \infty$ and hence $\omega(x)$ is empty. On the other hand, for $\dot{x} = ax$ the flow is $\varphi_t(x) = e^{at}x$, so that $\varphi_{t_k}(x) = e^{at_k}x \to 0$ as $t_k \to \infty$ if $a < 0$ giving $\omega(x) = \{0\}$ and clearly $\varphi_t(0) = 0$ so $\omega(x)$ is indeed invariant. But if $a > 0$ the set $\omega(x)$ is empty.

Example 2.12.

$$\begin{aligned}
\dot{x} &= x - y - x(x^2 + y^2), \\
\dot{y} &= x + y - y(x^2 + y^2).
\end{aligned} \tag{2.1}$$

By multiplying the first equation by x and the second by y and adding we obtain, after setting $r = \sqrt{x^2 + y^2}$ and simplifying, $\dot{r} = r - r^3$. The set $r = 1$, i.e. $\mathbb{S} = \{(x, y) : x^2 + y^2 = 1\}$, is an invariant set and $(x, y) = (0, 0)$ is the unique steady state. It is not difficult to see that any orbit is either the unique steady state $(0, 0)$, the unit circle, or a spiral that tends towards the unit circle. If $(x, y) \neq (0, 0)$, $\omega((x, y)) = \mathbb{S}$, and otherwise $\omega((0, 0)) = \{(0, 0)\}$.

Problem 2.13. *Find the omega limit sets for the predator–prey system on* $\mathbb{R}^2_{\geq 0}$

$$\begin{aligned}
\dot{x} &= x(1 - x + y), \\
\dot{y} &= y(-1 - y + x).
\end{aligned}$$

The many practical uses of the omega limit set is typified by the following result. Note that $\dot{x} = 1/x$ with $x(0) > 0$ satisfies $\dot{x} \to 0$ as $t \to \infty$, but the unique forward orbit $x(t) = \sqrt{2t + x(0)^2} \to \infty$ as $t \to \infty$ does not converge to a steady state. However, we do have the following lemma.

Lemma 2.14. *Suppose that $f : \mathbb{R}^n \to \mathbb{R}^n$ is continuously differentiable with isolated zeros. If $x : \mathbb{R}_{\geq 0} \to \mathbb{R}^n$ is a bounded forward orbit of $\dot{x} = f(x)$ such that $\dot{x}(t) \to 0$ as $t \to \infty$, then $x(t) \to p$ for some p as $t \to \infty$ where $f(p) = 0$, i.e. x converges to a steady state.*

Proof.   Let the orbit pass through x_0. $\overline{O^+(x_0)}$ is bounded and hence compact, so $\omega(x_0)$ is compact, connected and non-empty. For $p \in \omega(x_0)$ there exists a sequence $t_k \to \infty$ as $k \to \infty$ such that $x(t_k) \to p$ as $k \to \infty$. By continuity $0 = \lim_{k \to \infty} \dot{x}(t_k) = \lim_{k \to \infty} f(x(t_k)) = f(p)$, so that p is a steady state. Thus $\omega(x_0)$ consists entirely of steady states. Since $\omega(x_0)$ is connected, and the steady states are isolated, $\omega(x_0) = \{p\}$. $\square$

2.2.   *Stability*

Definition 2.15 (Lyapunov stability). A steady state x^* is said to be Lyapunov stable if for any $\epsilon > 0$ (arbitrarily small) $\exists \delta > 0$ such that $\forall x_0$ with $|x^* - x_0| < \delta$ we have $|\varphi(x_0, t) - x^*| < \epsilon$ for all $t \geq 0$.

A steady state is said to be unstable if it is not (Lyapunov) stable.

Definition 2.16 (Asymptotic stability). A steady state x^* is said to be asymptotically stable if it is Lyapunov stable and $\exists \rho > 0$ such that $\forall x_0$ with $|x^* - x_0| < \rho$ we have $|\varphi(x_0, t) - x^*| \to 0$ as $t \to \infty$.

For example, in the system $\dot{x} = -x - y + x(x^2 + y^2), \dot{y} = x - y + y(x^2 + y^2)$, the origin is locally asymptotically stable (we get $\dot{r} = -r + r^3$ by using polar coordinates). For a simple harmonic oscillator in the form of a pendulum, the pendulum resting vertically downwards is Lyapunov stable but not asymptotically stable unless there is damping such as air resistance. The upward vertical state of the pendulum is an example of an unstable steady state.

Definition 2.17 (Basin of attraction). The basin of attraction $B(x^*)$ of a steady state $x^* \in U$ is the set of points $y \in U$ such that $\varphi_t(y) \to x^*$ as $t \to \infty$.

Definition 2.18 (Global stability). If $B(x^*) = U$ then x^* is said to be globally asymptotically stable on U.

Problem 2.19. *Consider the logistic equation $\dot{x} = x(1-x)$. Find all forward invariant and invariant subsets of $\mathbb{R}_{\geq 0}$ and obtain the basin of attraction of the positive steady state.*

3. Ecological Systems

Consider the model

$$\dot{x}_i = x_i f_i(x), \quad i = 1, \ldots, n, \tag{3.1}$$

where each $f_i : \mathbb{R}^n \to \mathbb{R}^n$ is C^1. Suppose that $x(0) = (x_{01}, \ldots, x_{0n})$ has $x_{0k} = 0$ for $k \in J \subset \{1, \ldots, n\}$, so that some species are initially absent. Then these species are absent for all time for which the solutions exist.

Theorem 3.1. *For the model* (3.1) *the coordinate axes and the subspaces spanned by them, and $\mathbb{R}_{>0}^n$, are all forward invariant.*

In other words populations that start non-negative remain non-negative. Populations starting positive cannot go to zero in finite time.

4. LaSalle's Invariance Principle

We start with a basic result for Lyapunov functions (e.g. see p. 127 in Ref. 13).

Theorem 4.1. *Let $U \subseteq \mathbb{R}^n$ be open and $f : U \to \mathbb{R}$ be continuously differentiable and such that $f(x_0) = 0$ for some $x_0 \in U$. Suppose further that there is a real-valued function $V : U \to \mathbb{R}$ that satisfies* (i) $V(x_0) = 0$, (ii) $V(x) > 0$ *for $x \in U \setminus \{x_0\}$. Then if* (a) $\dot{V}(x) := \nabla V(x) \cdot f(x) \leq 0$ *for all $x \in U$, then x_0 is Lyapunov stable;* (b) *if*

$\dot{V}(x) < 0$ for all $U \setminus \{x_0\}$, then x_0 is asymptotically stable on U; (c) if $\dot{V}(x) > 0$ for all $x \in U \setminus \{x_0\}$, then x_0 is unstable.

This is a powerful theorem, but there is a useful generalisation of it which caters for when $\dot{V}^{-1}(0)$ is not an isolated point.

Theorem 4.2 (LaSalle's Invariance Principle). Let $\dot{x} = f(x)$ define a flow on a set $U \subseteq \mathbb{R}^n$, where f is continuously differentiable. Suppose $V : U \to \mathbb{R}$ is a continuously differentiable function. Let Q be the largest invariant subset of U. If for some bounded solution $x(t, x_0)$ with initial condition $x(0, x_0) = x_0 \in U$ the time derivative $\dot{V} = DVf$ satisfies $\dot{V}(x(t, x_0)) \leq 0$, then $\omega(x_0) \subseteq Q \cap \dot{V}^{-1}(0)$.

Proof. By boundedness of the orbit, $\omega(x)$ is non-empty and for $p \in \omega(x)$ there exists a $t_k \to \infty$ such that $x(t_k) \to p$. Since $\dot{V}(x(t_k)) \leq 0$, the sequence $\{V(t_k)\}$ is non-increasing. Since $x(t_k, x_0)$ is bounded, $V(x(t_k, x_0))$ is bounded, so that there exists a $c \in \mathbb{R}$ such that $V(t_k) \to c$. Hence $\omega(x) \subset V^{-1}(c)$. Since $\omega(x_0)$ is invariant, $\omega(x_0) \subset Q$, and for any $y \in \omega(x_0)$ we have $V(x(t, y)) = c$ and differentiating gives $\dot{V}(x(t, y)) = 0$ for all t, and hence $\dot{V}(y) = 0$ for all $y \in \omega(x)$. Hence $\omega(x) \subset Q \cap \dot{V}^{-1}(0)$. □

Example 4.3.

$$\dot{x} = x - y - x(x^2 + y^2),$$
$$\dot{y} = x + y - y(x^2 + y^2).$$

Taking $U = \mathbb{R}^2$, $V(x, y) = \sqrt{x^2 + y^2}$ we get

$$\frac{dV}{dt} = V(1 - V^2) \begin{cases} \leq 0 & \text{for } |(x, y)| \geq 1, \\ > 0 & |(x, y)| < 1. \end{cases}$$

Thus $\dot{V}^{-1}(0) = \{(0, 0)\} \cup \mathbb{S}$ ($\mathbb{S}$ is the unit circle). $Q = \mathbb{R}^2$ and applying LaSalle's Invariance Principle we get $\omega((x, y)) \subset \{(0, 0)\} \cup \mathbb{S}$. But the omega limit set is also connected, so that it must be either $\{(0, 0)\}$ or (by invariance) all of $\mathbb{S}$. Since $(0, 0)$ is unstable, we must have $\omega(x_0) = \mathbb{S}$ when $x_0 \neq 0$ and $\omega((0, 0)) = \{(0, 0)\}$.

Example 4.4.

$$\dot{x} = x(-\alpha + \gamma y), \tag{4.1}$$

$$\dot{y} = \alpha x - (\gamma x + \delta)y \quad (\alpha, \beta, \delta > 0). \tag{4.2}$$

This system has a unique steady state $(0,0)$, and one can show that $U = \mathbb{R}^2_{\geq 0}$ is forward invariant. Adding (4.1) and (4.2) we obtain

$$\frac{d}{dt}(x + y) = -\delta y \leq 0 \text{ on } \mathbb{R}^2_{\geq 0}.$$

Take $V(x, y) = x + y$. Then $\dot{V}^{-1}(0) = \{(s, 0) : s \in \mathbb{R}\}$. By LaSalle's Invariance Principle,

$$\omega((x, y)) \subseteq \{(s, 0) : s \in \mathbb{R}_{\geq 0}\}, \quad (x, y) \in \mathbb{R}^2_{\geq 0}.$$

But $\omega((x, y))$ must be connected and invariant, and the only invariant subsets of $T = \{(s, 0) : s \in \mathbb{R}_{> 0}\}$ for the flow of (4.1) and (4.2) are the origin and T itself. But, by (4.1), for $s \geq 0$, $\varphi_{t_k}(s, 0) \to (0, 0)$ for any sequence $t_k \to \infty$, so $\omega((x, y)) = \{(0, 0)\} \ \forall (x, y) \in \mathbb{R}^2_{\geq 0}$.

Theorem 4.5 (Goh[7]). *Suppose that the Lotka–Volterra system* $\dot{x}_i = x_i f_i(x) = x_i(r_i + \sum_{j=1}^n a_{ij} x_j)$, $i = 1, \ldots, n$, *has a unique interior steady state* $x^* = -A^{-1}r \in \mathbb{R}^n_{> 0}$. *Then this steady state is globally attracting on* $\mathbb{R}^n_{> 0}$ *if there exists a diagonal matrix* $D > 0$ *such that* $AD + DA^T$ *is negative definite.*

Proof. Let $V : \mathbb{R}^n_{\geq 0} \to \mathbb{R}_{\geq 0}$ be defined by

$$V(x) = \sum_{i=1}^n \alpha_i \left(x_i - x_i^* - x_i^* \log(x_i/x_i^*) \right),$$

where $\alpha_i \in \mathbb{R}_{> 0}$ are to be found. Then we compute

$$\dot{V} = \nabla V \cdot f$$

$$= \sum_{i=1}^n \alpha_i (x_i - x_i^*) f_i(x) = \sum_{i=1}^n \alpha_i (x_i - x_i^*) \left\{ \sum_{j=1}^n a_{ij} (x_j - x_j^*) \right\}.$$

This can be rewritten as

$$\dot{V} = (x - x^*)^T A^T D (x - x^*) = \frac{1}{2}(x - x^*)^T (DA + A^T D)(x - x^*),$$

where $D = \text{diag}(\alpha_1, \ldots, \alpha_n)$. When $DA + A^T D$ is negative definite, $\dot{V} \leq 0$ and $\dot{V}^{-1}(0) = \{x^*\}$. V is convex (as the sum of convex functions) and has a unique minimum at $x = x^*$. Hence, by Theorem 4.1, x^* is globally asymptotically stable on $\mathbb{R}^n_{>0}$. □

(See Ref. 16 for an improvement of this result to cater for boundary steady states.)

Example 4.6. Consider the two species Lotka–Volterra system

$$\begin{aligned} \dot{x} &= x(a + bx + cy), \\ \dot{y} &= y(d + ex + fy). \end{aligned} \qquad (4.3)$$

Suppose that (4.3) has a unique interior steady state, say $(x^*, y^*) \in \mathbb{R}^2_{>0}$. Thus $bf - ce \neq 0$. We use Theorem 4.5. Let $\lambda > 0$ and

$$D = \begin{pmatrix} 1 & 0 \\ 0 & \lambda \end{pmatrix}, \quad M = DA + A^T D = \begin{pmatrix} 2b & c + \lambda e \\ c + \lambda e & 2\lambda f \end{pmatrix}.$$

Then for diagonal stability we need M to be negative definite, that is, if and only if its trace is negative and its determinant is positive,

(i) $\lambda f + b < 0$, (ii) $4bf\lambda > (c + \lambda e)^2$.

Since we seek $\lambda > 0$, to satisfy (ii) we require $fb > 0$, which then implies $f, b < 0$ by (i). Next for (ii) we need

$$4bf\lambda - (c + \lambda e)^2 = (4bf - 2ce)\lambda - c^2 - e^2\lambda^2 > 0$$

for some $\lambda > 0$. The quadratic $\phi(\lambda) = (4bf - 2ce)\lambda - c^2 - e^2\lambda^2$ is negative for $\lambda = 0$ and large $|\lambda|$, and so is positive for some $\lambda > 0$ only if $4bf - 2ec = 2 \det A + 2bf > 0$ and $(4bf - 2ce)^2 > 4e^2c^2$ which simplifies to $\det A > 0$.

To conclude, we have shown the following theorems.

Theorem 4.7 (Goh[6]). *Suppose the system*

$$\begin{aligned} \dot{x} &= x(a + bx + cy), \\ \dot{y} &= y(d + ex + fy). \end{aligned} \qquad (4.4)$$

has a unique interior steady state $(x^*, y^*) \in \mathbb{R}^2_{>0}$. Then (x^*, y^*) globally attracts all points in $\mathbb{R}^2_{>0}$ if $f < 0, b < 0$ and $\det A > 0$.

Problem 4.8. *Is the converse of Theorem 4.7 true?*

5. Conservative Lotka–Volterra Systems

Definition 5.1 (Conservative Lotka–Volterra). We will say that (1.1) is conservative if there exists a diagonal matrix $D > 0$ such that AD is skew-symmetric.

Notice that if B is skew-symmetric then $b_{ij} = -b_{ji}$ for all i, j. In particular $b_{ii} = -b_{ii}$ so that $b_{ii} = 0$, i.e. the diagonal elements of a skew-symmetric matrix are all zero.

Problem 5.2. *Consider the two-species Lotka–Volterra system*

$$\frac{1}{N}\frac{dN}{dt} = a - bP,$$

$$\frac{1}{P}\frac{dP}{dt} = cN - d.$$

Change to new coordinates $x = \log N, y = \log P$ *and show that* $H(x, y) = dx + ay - e^x - e^y$ *is constant along a trajectory* $(x(t), y(t))$. *Show also that* $\dot{x} = \frac{\partial H}{\partial y}, \dot{y} = -\frac{\partial H}{\partial x}$.

A change of coordinates $y_i = x_i/d_i$ $(d_i \leq 0)$ transforms (1.1) into

$$\dot{y}_i = y_i\left(r_i + \sum_{j=1}^{n} d_j a_{ij} y_j \right),$$

so that we obtain another Lotka–Volterra system with interaction matrix AD. The Lotka–Volterra systems with interaction matrices AD for $D > 0$ diagonal have topologically equivalent dynamics.

Lemma 5.3. *If A is an $n \times n$ skew-symmetric matrix, then $\det A = (-1)^n \det A$. Hence when n is odd, A is singular.*

Proof. $\det A = \det A^T = \det(-A) = (-1)^n \det A.$ □

Now suppose that A is skew-symmetric. We will show that certain Lotka–Volterra systems can be written in Hamiltonian form. But before doing so, we recall the definition of a Hamiltonian system on $\mathbb{R}^n$ (see, for example, Ref. 12). Let C^∞ denote the space of smooth functions $\mathbb{R}^n \to \mathbb{R}$.

Definition 5.4 (Hamiltonian system on $\mathbb{R}^n$). A Hamiltonian system (on $\mathbb{R}^n$) is a pair $(H, \{\cdot, \cdot\})$ where $H : \mathbb{R}^n \to \mathbb{R}$ is a smooth function, called the Hamiltonian, and $\{\cdot, \cdot\} : C^\infty \times C^\infty \to C^\infty$ is a Poisson bracket; that is a bilinear skew-symmetric map $\{\cdot, \cdot\} : C^\infty \times C^\infty \to C^\infty$ that satisfies the following relations for all $f, g, h \in C^\infty$:

(1) $\{f, gh\} = \{f, g\}h + g\{f, h\}$ [Leibnitz rule];
(2) $\{f, \{g, h\}\} + \{g, \{h, f\}\} + \{h, \{f, g\}\} = 0$ [Jacobi Identity].

For example, when $n = 2$ the bracket $\{\cdot, \cdot\} : C^\infty \times C^\infty \to \mathbb{R}$ given by

$$\{f, g\} = \frac{\partial f}{\partial q}\frac{\partial g}{\partial p} - \frac{\partial f}{\partial p}\frac{\partial g}{\partial q}$$

defines a Poisson bracket.

For each $g \in C^\infty$, the bracket defines a Hamiltonian vector field X_g on $\mathbb{R}^n$ via $\{f, g\} = X_g(f)$. In the previous example $X_g = \frac{\partial g}{\partial q}\frac{\partial}{\partial p} - \frac{\partial g}{\partial p}\frac{\partial}{\partial q}$. Hamilton's equations are then given by $\dot{x}_i = X_H(x_i)$ for $i = 1, \ldots, n$. In particular, $\dot{H} = \{H, H\} = 0$ gives the constancy of the Hamiltonian function along an orbit. In addition to conserved functions conserved on orbits, there may also be functions C such that $\{C, f\} = 0$ for all functions $f \in C^\infty$. That is, C is constant along all flows generated by the Hamiltonian vector fields X_f as f ranges through C^∞. Such functions C are known as *Casimirs*.

To establish that a Lotka–Volterra system is Hamiltonian, we thus have to identify both a Poisson bracket and a Hamiltonian function.

Before turning to a Hamiltonian description of (1.1) we note that there is a graphical way to test whether a Lotka–Volterra system is conservative.

Proposition 5.5 (Volterra[18]). *The Lotka–Volterra system $\dot{x}_i = x_i(r_i + (Ax)_i)$ is conservative if and only if $a_{ii} = 0$ and $a_{ij} \neq 0 \Rightarrow a_{ij}a_{ji} < 0$, and for every sequence $i_1, i_2, \ldots, i_s$ we have $a_{i_1 i_2} a_{i_2 i_3} \cdots a_{i_s i_1} = (-1)^s a_{i_s i_{s-1}} \cdots a_{i_2 i_1} a_{i_1 i_s}$.*

That is we have a graphical condition that there exists a diagonal matrix $D > 0$ such that AD is skew-symmetric ($AD + DA^T = 0$; compare with Theorem 4.5). One creates a signed digraph with nodes labelled 1 to n where n is the number of species and puts on each directed edge linking nodes i to j the number a_{ij}. The condition to check is then that for each cycle in the digraph of length s, the product of the edge numbers in one direction is $(-1)^s$ times the product in the opposite direction.

6. Volterra's Construction[5,18]

We start with the skew-symmetric system

$$\dot{x}_i = x_i \left(r_i + \sum_{j=1}^{n} a_{ij} x_j \right), \quad a_{ij} = -a_{ji}. \tag{6.1}$$

Volterra introduced new coordinates which he called *quantity of life*:

$$Q_i = \int_0^t x_i(s)\, ds \quad (i = 1, \ldots, n).$$

Thus $\dot{Q}_i = x_i$ and (6.1) becomes the second-order system

$$\ddot{Q}_i = \dot{Q}_i \left(r_i + \sum_{j=1}^{n} a_{ij} Q_j \right). \tag{6.2}$$

Then he introduces $H(Q, \dot{Q}) = \sum_{i=1}^{n}(r_i Q_i - \dot{Q}_i)$ so that

$$\frac{dH}{dt} = \sum_{i=1}^{n}(r_i \dot{Q}_i - \ddot{Q}_i)$$

$$= \sum_{i=1}^{n} \left(r_i \dot{Q}_i - \dot{Q}_i \left(r_i + \sum_{j=1}^{n} a_{ij} \dot{Q}_j \right) \right) = - \sum_{i,j=1}^{n} a_{ij} \dot{Q}_i \dot{Q}_j = 0$$

using skew-symmetry of $A = ((a_{ij}))$. Dual variables P_i are then defined via

$$P_i = \log \dot{Q}_i - \frac{1}{2} \sum_{j=1}^{n} a_{ij} Q_j \quad (i = 1, \ldots, n).$$

In terms of these new coordinates, we get the transformed $h(Q, P) = H(Q, \dot{Q})$ where

$$h(Q, P) = \sum_{i=1}^{n} \left(r_i Q_i - \exp\left(P_i + \frac{1}{2} \sum_{j=1}^{n} a_{ij} Q_j \right) \right).$$

Now we can check that

$$\frac{dQ_i}{dt} = \exp\left(P_i + \frac{1}{2} \sum_{j=1}^{n} a_{ij} Q_j \right) = -\frac{\partial h}{\partial P_i},$$

and

$$\begin{aligned}
\frac{dP_i}{dt} &= \frac{d}{dt} \left\{ \log \dot{Q}_i - \frac{1}{2} \sum_{j=1}^{n} a_{ij} Q_j \right\} \\
&= \frac{\ddot{Q}_i}{\dot{Q}_i} - \frac{1}{2} \sum_{j=1}^{n} a_{ij} \dot{Q}_j = r_i + \sum_{j=1}^{n} a_{ij} \dot{Q}_j - \frac{1}{2} \sum_{j=1}^{n} a_{ij} \dot{Q}_j \\
&= r_i + \frac{1}{2} \sum_{j=1}^{n} a_{ij} \exp\left(P_j + \frac{1}{2} \sum_{k=1}^{n} a_{jk} Q_k \right).
\end{aligned}$$

On the other hand

$$\begin{aligned}
\frac{\partial h}{\partial Q_i} &= r_i - \sum_{k=1}^{n} \frac{a_{ki}}{2} \exp\left(P_k + \frac{1}{2} \sum_{j=1}^{n} a_{kj} Q_j \right) \\
&= r_i + \sum_{k=1}^{n} \frac{a_{ik}}{2} \exp\left(P_k + \frac{1}{2} \sum_{j=1}^{n} a_{kj} Q_j \right),
\end{aligned}$$

using $a_{ik} = -a_{ki}$. This gives $\dot{P}_i = \frac{\partial h}{\partial Q_i}$ as required.

Hence we have shown that the system (6.1) is canonically Hamiltonian in the new coordinates P, Q with Hamiltonian function

$$h(P, Q) = \sum_{i=1}^{n} \left(r_i Q_i - \exp(P_i + \frac{1}{2} \sum_{j=1}^{n} a_{ij} Q_j) \right),$$

and the standard Poisson bracket

$$\{f, g\} = \sum_{i=1}^{n} \frac{\partial f}{\partial P_i} \frac{\partial g}{\partial Q_i} - \frac{\partial g}{\partial Q_i} \frac{\partial f}{\partial P_i}.$$

7. An Alternative Hamiltonian Formulation

In the previous formulation, we doubled the number of variables in order to find a Hamiltonian structure. Here we keep the same number of variables as the original Lotka–Volterra system.

Suppose that $Ax + r = 0$ has a solution $x^* \in \mathbb{R}^n$ (here A is skew-symmetric). Introduce new variables $y_i = \log x_i$:

$$\dot{y}_i = \left(r_i + \sum_{j=1}^{n} a_{ij} \exp y_j \right) = \sum_{j=1}^{n} a_{ij} (\exp y_j - x_j^*).$$

Now define

$$H(y) = \sum_{i=1}^{n} (\exp y_i - x_i^* y_i),$$

so that

$$\dot{y}_i = \sum_{j=1}^{n} a_{ij} (e^{y_j} - x_j^*) = \sum_{j=1}^{n} a_{ij} \frac{\partial H}{\partial y_j}, \tag{7.1}$$

$$\frac{dH}{dt} = \sum_{j=1}^{n} \frac{\partial H}{\partial y_j} \dot{y}_j$$

$$= \sum_{i=1}^{n} \sum_{j=1}^{n} a_{ij} \frac{\partial H}{\partial y_i} \frac{\partial H}{\partial y_j}$$

$$= \frac{1}{2} \sum_{i=1}^{n} \sum_{j=1}^{n} (a_{ij} + a_{ji}) \frac{\partial H}{\partial y_i} \frac{\partial H}{\partial y_j}$$

$$= 0,$$

using skew-symmetry of $A = ((a_{ij}))$. To complete the Hamiltonian formulation we check that

$$\{f, g\} = \nabla f \cdot A \nabla g \tag{7.2}$$

provides a suitable Poisson bracket.

Problem 7.1. *Show that (7.2) defines a Poisson bracket.*

Notice that the x^* need not lie in the first quadrant. In an odd-dimensional Lotka–Volterra system with skew-symmetric interaction matrix A, we have $\det A = 0$ and it is possible that $Ax + r = 0$ has no solutions. Indeed, if A is singular, then there is a $v \neq 0$ in $\ker A$ such that $v^T A = (A^T v)^T = -(Av)^T = 0$. Thus for a solution to exist we must have $v^T r = 0$ for all $v \in \ker A$, i.e. $r \in (\ker A)^{\perp}$.

Example 7.2. Consider the Lotka–Volterra system for three interacting species:

$$\dot{x}_1 = x_1(r_1 + w_1 x_2 - w_2 x_3),$$
$$\dot{x}_2 = x_2(r_2 - w_1 x_1 + w_3 x_3), \qquad (7.3)$$
$$\dot{x}_3 = x_3(r_3 + w_2 x_1 - w_3 x_2),$$

where $w_1, w_2, w_3 > 0$ and each $r_i > 0$. Here species 3 is prey to species 2. Species 2 consumes species 3, but is consumed by species 1. Species 1 consumes species 2, but it is consumed by species 3. (So we have a cycle of interactions.) It is easy to see that the interaction matrix

$$A = \begin{pmatrix} 0 & w_1 & -w_2 \\ -w_1 & 0 & w_3 \\ w_2 & -w_3 & 0 \end{pmatrix}$$

is skew-symmetric. Since A is 3×3, we already know that A is singular. Thus if q is a solution to $Aq + r = 0$ then so too is $q + k$ for any $k \in \ker A = \{\alpha(w_3, w_2, w_1) : \alpha \in \mathbb{R}\}$. One finds that $Aq + r = 0$ has no solutions (in $\mathbb{R}^3$) unless

$$v^T r = w_3 r_1 + w_1 r_3 + w_2 r_2 = 0 \qquad (7.4)$$

($v = (w_3, w_2, w_1)$) and in this case $q = (\frac{r_2}{w_1}, -\frac{r_1}{w_1}, 0) + \alpha v$ for $\alpha \in \mathbb{R}$.

Thus let us now assume that (7.4) holds. For the Hamiltonian we may take

$$H(x) = x_1 + x_2 + x_3 - \frac{r_2}{w_1} \log x_1 + \frac{r_1}{w_1} \log x_2.$$

We find that

$$\dot{H} = \left(r_3 + \frac{r_2\omega_2}{\omega_1} + \frac{r_1\omega_3}{\omega_1} \right) x_3 = 0$$

by virtue of (7.4). A suitable Poisson bracket is thus

$$\{f, g\} = \omega_1 x_1 x_2 \left(\frac{\partial f}{\partial x_1} \frac{\partial g}{\partial x_2} - \frac{\partial g}{\partial x_1} \frac{\partial f}{\partial x_2} \right)$$
$$- \omega_2 x_1 x_3 \left(\frac{\partial f}{\partial x_1} \frac{\partial g}{\partial x_3} - \frac{\partial g}{\partial x_1} \frac{\partial f}{\partial x_3} \right)$$
$$+ \omega_3 x_2 x_3 \left(\frac{\partial f}{\partial x_2} \frac{\partial g}{\partial x_3} - \frac{\partial g}{\partial x_2} \frac{\partial f}{\partial x_3} \right).$$

Since A is singular, there are Casimir functions C; that is C satisfying $\{C, g\} = 0$ for all g, proportional to

$$C(x) = \omega_3 \log x_1 + \omega_2 \log x_2 + \omega_1 \log x_3$$

(or we could take $C(x) = x_1^{\omega_3} x_2^{\omega_2} x_3^{\omega_1}$). We find that

$$\dot{C} = r_3\omega_1 + r_2\omega_2 + r_1\omega_3 = 0,$$

again using (7.4). The dynamics lies on the intersection of the surfaces $H(x) = H(x(0))$ and $C(x) = C(x(0))$ in the first quadrant.

Let us change coordinates, setting $X = \log x_1$, $Y = \log x_2$ and $Z = \log x_3$. Then we have on a solution

$$e^X + e^Y + e^Z - \frac{r_2}{\omega_1} X + \frac{r_1}{\omega_1} Y = A,$$
$$\omega_3 X + \omega_2 Y + \omega_1 Z = B,$$

where A, B are constants. Hence we may plot

$$Z = \log \left(A - e^X - e^Y + \frac{r_2}{\omega_1} X - \frac{r_1}{\omega_1} Y \right), \qquad (7.5)$$

$$Z = \frac{B - \omega_3 X - \omega_2 Y}{\omega_1}. \qquad (7.6)$$

The first surface is concave where the logarithm is defined. Searching for periodic orbits then becomes the study of how the surface (7.5)

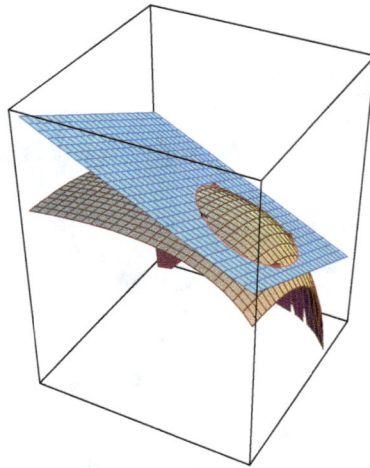

Fig. 1. A periodic solution to the three species model (7.3). There exists a continuum of periodic orbits around the interior steady state.

intersects the plane (7.6). An example a periodic orbit is shown in Fig. 1.

8. Cooperative Lotka–Volterra Systems

We will consider the general Lotka–Volterra system

$$\dot{x}_i = F_i(x) := x_i \left(r_i + \sum_{j=1}^{n} a_{ij} x_j \right), \quad (i = 1, \ldots, n). \tag{8.1}$$

except that we will constrain ourselves to the case that $a_{ij} \geq 0$ when $i \neq j$, i.e. the off-diagonal elements of the interaction matrix are non-negative. Notice that in this case

$$\frac{\partial F_i}{\partial x_j} = a_{ij} x_i \geq 0, \quad i \neq j,$$

since for $i \neq j$ we have $a_{ij} \geq 0$ and we have $x \in \mathbb{R}^n_{\geq 0}$. Since the first quadrant is invariant, the Jacobian has non-negative off-diagonal elements.

Definition 8.1 (Cooperative matrix). We will say that any real $n \times n$ matrix with non-negative off-diagonal elements is *cooperative*.

Some notation

In what follows we will use the following notation for ordering vectors $x \in \mathbb{R}^n$: For each $x, y \in \mathbb{R}^n$

- $x \le y \Leftrightarrow x_i \le y_i$ for all $i = 1, \ldots, n$;
- $x < y \Leftrightarrow x_i \le y_i$ for all $i = 1, \ldots, n$, but $x_k \ne y_k$ for some k;
- $x \ll y \Leftrightarrow x_i < y_i$ for all $i = 1, \ldots, n$.

(Similarly for $\ge, >, \gg$.) We say that $(\mathbb{R}^n, \le)$ is an ordered vector space.

The following Perron–Frobenius theorem is fundamental in the study of coorperative or competitive systems (see, for example, Ref. 3). We recall that the spectral radius of A, written $\rho(A)$, is the modulus of an eigenvalue of A of largest modulus, and a matrix A is irreducible if it is not similar via a permutation to a block upper triangular matrix (that has more than one block of positive size).

Theorem 8.2 (Perron–Frobenius). *If A is an $n \times n$ real matrix with non-negative entries. Then*

- *$\rho(A)$ is an eigenvalue of A;*
- *A has left and right eigenvectors $u > 0$ and $v > 0$ associated with $\rho(A)$ (i.e. $uA = \rho(A)u$ and $Av = \rho(A)v$).*

If A is also irreducible then we have $\rho(A) > |\mu|$ for any eigenvalue $\mu \ne \rho(A)$, and $\rho(A)$ is simple and $v \gg 0$ and $u \gg 0$ in the above statements.

(*Note*: the inequalities in Theorem 8.2 use the vector ordering defined above).

Definition 8.3. A matrix A is negatively (row) diagonally dominant if there exists a $d \gg 0$ such that $a_{ii}d_i + \sum_{j \ne i} |a_{ij}|d_j < 0$ for all $i = 1, \ldots, n$.

When A is a cooperative matrix this becomes $Ad \ll 0$.

Lemma 8.4. *Let A be a cooperative matrix. Then A is stable if and only if it is negatively diagonally dominant.*

Proof. First suppose that A is negatively diagonally dominant: There exists a $d \gg 0$ such that $Ad \ll 0$. Note that we must have all $a_{ii} < 0$ since the off-diagonal elements are non-negative and $d \gg 0$. Let λ be an eigenvalue of A with right eigenvector x. Let $y_i = x_i/d_i$ for $i = 1, \ldots, n$ and $|y_m| = \max_i |y_i| > 0$. Then $\lambda d_i y_i = \sum_{j=1}^n a_{ij} d_j y_j$ and

$$\lambda d_m = d_m a_{mm} + \sum_{j \neq m}^n d_j a_{mj} \frac{y_j}{y_m}.$$

Therefore

$$|\lambda d_m - d_m a_{mm}| \leq \sum_{j \neq m}^n d_j a_{mj} \left| \frac{y_j}{y_m} \right| \leq \sum_{j \neq m}^n d_j a_{mj} < -d_m a_{mm}$$

by hypothesis. Hence $|\lambda - a_{mm}| < -a_{mm}$ and λ must lie in the open disc in the Argand plane whose boundary passes through zero and whose centre is at the negative number a_{mm}. Thus all eigenvalues λ have negative real part, so A is stable.

Conversely, suppose that A is stable and has non-negative off-diagonal elements. For $c > 0$ sufficiently large $B = A + cI$ is a non-negative matrix and so by the Perron–Frobenius theorem there is a $\lambda = \rho(B) \geq 0$ and a $v > 0$ such that $Bv = \lambda v = \rho(B)v$. But then $Av = (\rho(B) - c)v$ so that, since A is stable, $\rho(B) < c$. Since $\rho(B) < c$ the following series converges

$$A^{-1} = -\frac{1}{c}\left(I + \frac{1}{c}B + \frac{1}{c^2}B^2 + \cdots\right)$$

and thus all elements of A^{-1} are non-positive. Now set $d = -A^{-1}(1, \ldots, 1)^T$. Then $d \gg 0$ (no row of A can be zero, since it is non-singular) and $Ad = -(1, \ldots, 1)^T \ll 0$. □

Corollary 8.5. *If A is cooperative and $r \gg 0$ then $Ax + r = 0$ has a unique interior solution $x \in \mathbb{R}^n_{>0}$ if and only if A is stable.*

Problem 8.6. *For the system (4.4) when $c > 0$, $e > 0$ and $\det A > 0$, find the condition for a unique interior steady state.*

We also have the following (see, for example, Theorem 15.1.1 in Ref. 10).

Theorem 8.7 (Global convergence for cooperative Lotka–Volterra). *When A cooperative and stable, the system* (8.1) *has a unique interior steady state that attracts* $\mathbb{R}^n_{>0}$.

Proof. By Corollary 8.5, A is negatively diagonally dominant by Lemma 8.4, i.e there exists a $d \gg 0$ such that $a_{ii}d_i + \sum_{j=1}^n a_{ij}d_j < 0$. Define

$$V(x) = \max_k \frac{|x_k - x^*_k|}{d_k}.$$

Then $V(x) \geq 0$ with equality if and only if $x = x^*$. Now, consider a time interval during which $\max_k \frac{|x_k - x^*_k|}{d_k} = \frac{|x_i - x^*_i|}{d_i}$. Then

$$\dot{V} = \frac{1}{d_i} \dot{x}_i \operatorname{sgn}(x_i - x^*_i)$$

$$= \frac{x_i}{d_i} \left\{ a_{ii}(x_i - x^*_i) + \sum_{j \neq i} a_{ij}(x_j - x^*_j) \right\} \operatorname{sgn}(x_i - x^*_i)$$

$$\leq \frac{x_i}{d_i} \left\{ a_{ii}|x_i - x^*_i| + \sum_{j \neq i} a_{ij}|x_j - x^*_j| \right\}$$

$$\leq \frac{x_i}{d_i} V(x) \left\{ a_{ii}d_i + \sum_{j \neq i} a_{ij}d_j \right\}$$

$$\leq 0 \quad \text{for all } x \in \mathbb{R}^n_{>0}, \text{ with equality if and only if } x = x^*.$$

Hence by Theorem 4.1, $x(t) \to x^*$ as $t \to \infty$. □

9. Competitive Lotka–Volterra Systems

Now we consider the Lotka–Volterra system

$$\dot{x}_i = x_i\left(r_i - \sum_{j=1}^n a_{ij}x_j\right) = F_i(x), \quad i = 1, \ldots, n, \tag{9.1}$$

under the special conditions that $a_{ij} > 0$ for all $1 \leq i, j \leq n$ (**caution:** notice the change of sign in (9.1)). This means that each species competes with all other species including itself. If some $r_i \leq 0$ then it is clear that $x_i(t) \to 0$ as $t \to \infty$ since $\mathbb{R}^n_{\geq 0}$ is invariant and

$$\dot{x}_i = x_i \left(r_i - \sum_{j=1}^n a_{ij} x_j \right) \leq -a_{ii} x_i^2 \leq 0,$$

with equality if and only if $x_i = 0$. We will therefore also assume $r_i > 0$ for each $i = 1, \ldots, n$. This means that in the absence of any competitors the species i will evolve according to $\dot{x}_i = x_i(r_i - a_{ii}x_i)$ and hence will either remain at zero or stabilise at its carrying capacity $K_i = r_i/a_{ii} > 0$. It also means that the origin is an unstable node.

Lemma 9.1. *Since $a_{ij} > 0$ and $r_i > 0$, all orbits of (9.1) are bounded.*

Proof. $\mathbb{R}^n_{\geq 0}$ is invariant and

$$\dot{x}_i = r_i x_i - x_i \sum_{j=1}^n a_{ij} x_j \leq r_i x_i - a_{ii} x_i^2$$

$$= x_i(r_i - a_{ii}x_i) < 0 \text{ if } x_i > \frac{r_i}{a_{ii}},$$

so that the ith species is bounded for each $i = 1, \ldots, n$. □

10.   Smale's Construction

In the 1970s many thought that for a finite habitat that is home to a number of species that compete with each other and the other species, the long-term outcome is "simple" dynamics, e.g. convergence to a steady state or a periodic orbit. But this is not the case, as Stephen Smale showed in 1976.[14] Consider a more general model of **total competition**:

$$\dot{x}_i = x_i M_i(x) = F_i(x), \quad (i = 1, \ldots, n), \tag{10.1}$$

where M_i is smooth and we will suppose that

S1: For all pairs i, j we have $\frac{\partial M_i}{\partial x_j} < 0$ when $x_i > 0$ (totally competitive).

S2: There is a constant K such that for each i, $M_i(x) < 0$ if $|x| > K$.

Condition S1 means that

$$\frac{\partial \dot{x}_i}{\partial x_j} = x_i \frac{\partial M_i}{\partial x_j} < 0 \quad \text{for all } i, j \text{ if } x_i > 0. \tag{10.2}$$

Thus the Jacobian has negative elements in $\mathbb{R}^n_{>0}$; in other words competition for resources. The second condition says that there are finite resources and that the populations cannot grow indefinitely.

Smale showed that examples of systems satisfying (10.1) and the conditions S1, S2 whose long-term dynamics lie on a simplex and obey $\dot{x} = h(x)$ on the simplex, where h is any smooth vector field of our choice! Thus the simplex is an attractor upon which arbitrary dynamics can be specified, even chaos.

We follow the presentation of Ref. 9. Let $\Delta_1 = \{x \in \mathbb{R}^n_{\geq 0} : \|x\|_1 = 1\}$ be the standard probability simplex with tangent space $\Delta_0 = \{x \in \mathbb{R}^n : \sum_{i=1}^n x_i = 0\}$. Let $h_0 : \Delta_1 \to \Delta_0$ be a smooth vector field on Δ_1 whose components can be written as $h_{0i}(x) = x_i g_i(x)$ and $h : \mathbb{R}^n_{\geq 0} \to \Delta_0$ any smooth map which agrees with h_0 on Δ_1.

Now let $\beta : \mathbb{R} \to \mathbb{R}$ be any smooth function which is 1 in a neighbourhood of 1 and $\beta(t) = 0$ if $t \leq \frac{1}{2}$ or $t \geq \frac{3}{2}$. For $\epsilon > 0$ define M_i on $\mathbb{R}^n_{\geq 0}$ by

$$M_i(x) = 1 - \|x\|_1 + \epsilon \beta(\|x\|_1) g_i(x), \quad 1 \leq i \leq n.$$

We may check: for each i, j,

$$\frac{\partial M_i}{\partial x_j} = -1 + \epsilon \beta'(\|x\|_1) g_i - \epsilon \beta(\|x\|_1) \frac{\partial g_i}{\partial x_j} < 0,$$

for small enough ϵ since β has compact support.

Now as before, $\mathbb{R}^n_{\geq 0}$ is invariant, and $\frac{d}{dt}\|x\|_1 = \sum_{i=1}^n \dot{x} = \|x\|_1(1 - \|x\|_1)$ (the logistic equation!). Thus Δ_1 is forward invariant and any point in $\mathbb{R}^n_{\geq 0} \setminus \{0\}$ is attracted to Δ_1. On Δ_1 we have

$$M_i(x) = 1 - \|x\|_1 + \epsilon \beta(\|x\|_1) g_i(x) = \epsilon g_i(x),$$

so that the dynamics on the attractor is $\dot{x}_i = x_i \epsilon g_i(x) = \epsilon h_i(x)$ for $i = 1, \ldots, n$, with h arbitrary.

Hence we should be warned that the long-term dynamics of bounded competitive systems in dimensions higher than two can be very complex (Although one can show [see the next section on the carrying simplex] that when $n = 3$ the long-term dynamics must lie on a set of dimension at most 2, and this severely restricts the possibilities. However, much more is possible when $n \geq 4$.)

11.   Carrying Simplices

A bounded totally competitive system with the origin unstable has a unique invariant manifold that attracts the first orthant minus the origin. We will give an example[a] of such a system where the invariant manifold can be explicitly found — it is the probability simplex in $\mathbb{R}^n_{\geq 0}$ — and all orbits save the origin are attracted to it. Moreover, (for that example) the dynamics on the simplex is canonically Hamiltonian and all orbits are periodic.

We consider again the system
$$\dot{x}_i = x_i M_i(x), \quad (i = 1, \ldots, n),$$
where M_i is smooth and we will suppose that

S1: For all pairs i, j we have $\frac{\partial M_i}{\partial x_j} < 0$.
S2: There is a constant K such that for each i, $M_i(x) < 0$ if $|x| > K$.
S3: $M_i(0) > 0$.

Condition S3 makes the origin 0 a repelling steady state. Since orbits are bounded, the basin of repulsion of 0 in $\mathbb{R}^n_{\geq 0}$ is bounded. The boundary of the basin of repulsion is called the *Carrying Simplex* and is denoted by Σ. One can think of Σ as being the boundary of the set of points whose α limit is the origin.

All steady states and all ω limit sets lie in Σ and we have from Ref. 8.

Theorem 11.1 (The Carrying Simplex). *Given* (10.1) *every trajectory in* $\mathbb{R}^n_{\geq 0} \setminus \{0\}$ *is asymptotic to one in* Σ, *and* Σ *is a Lipschitz submanifold, everywhere transverse to all strictly positive directions, and homeomorphic to the probability simplex.*

Thus totally competitive n−dimensional Lotka–Volterra systems (as above) eventually evolve like $(n - 1)$-dimensional systems. Thus

---

[a] A second example, since in Smale's example the unit simplex is also a carrying simplex.

nothing very exotic can happen for $n < 4$. In Fig. 3 we display three examples of the carrying simplex for totally competitive Lotka–Volterra systems.[1]

The following example has the advantage that the carrying simplex can be found explicitly, and it is easy to see that all points save the origin are attracted to it (Fig. 2).

Example 11.2. We consider the illustrative example of an eventually periodic competitive system[11]

$$\dot{x} = x(1 - x - \alpha y - \beta z),$$

$$\dot{y} = y(1 - \beta x - y - \alpha z),$$

$$\dot{z} = z(1 - \alpha x - \beta y - z),$$

where $\alpha, \beta > 0$ and $\alpha + \beta = 2$. Let

$$V(x, y, z) = xyz.$$

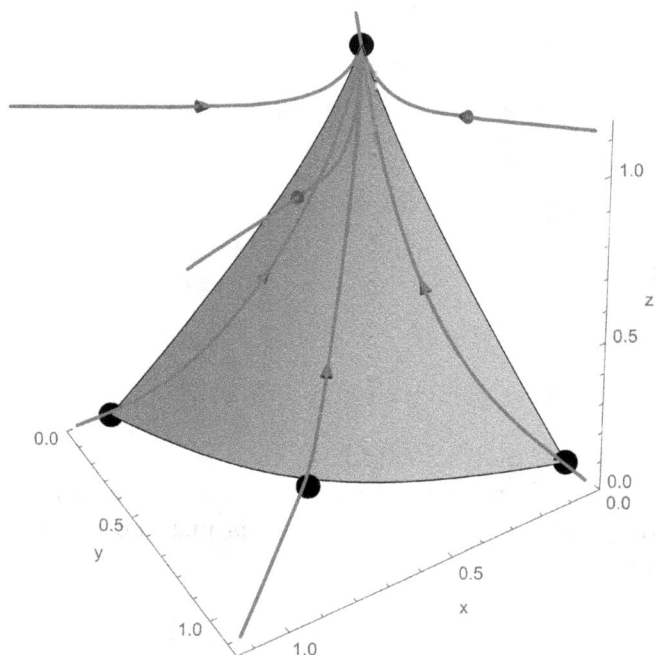

Fig. 2. The Carrying Simplex attracts all orbits except the origin and contains any ω limit set and in particular all steady states except the origin.

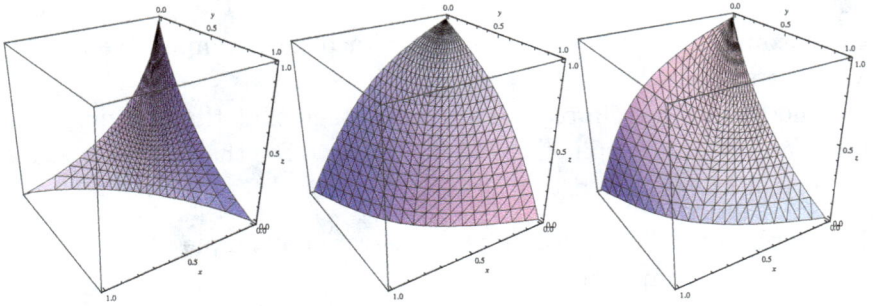

Fig. 3.   Examples of the carrying simplex for competitive the three-dimensional Lotka–Volterra equations. From left to right the carrying simplex is (i) convex, (ii) concave and (iii) saddle-like.[1]

Then

$$\frac{d}{dt}V = xyz\left(\frac{\dot{x}}{x} + \frac{\dot{y}}{y} + \frac{\dot{z}}{z}\right)$$

$$= V\left((1 - x - \alpha y - \beta z)\right.$$

$$+ (1 - \beta x - y - \alpha z) + (1 - \alpha x - \beta y - z))$$

$$= V(3 - (x + y + z) - (\alpha + \beta)(x + y + z))$$

$$= 3V(1 - (x + y + z)) \quad \text{since } \alpha + \beta = 2.$$

Moreover

$$\frac{d}{dt}(x + y + z) = (x + y + z) - x^2 - y^2 - z^2 - (\alpha + \beta)(xy + xz + yz)$$

$$= (x + y + z)(1 - (x + y + z)).$$

Thus if $(x_0, y_0, z_0) \in \mathbb{R}^3 \setminus (0, 0, 0)$ we have $x(t) + y(t) + z(t) \to 1$ as $t \to \infty$. That is, all orbits eventually end up on the simplex Δ_1. Thus the carrying simplex Σ in this example is just the simplex Δ_1. On Δ_1 we have

$$\frac{dV}{dt} = 3V(1 - (x + y + z)) = 0,$$

that is $V = \text{const}$ on Δ_1. What is the dynamics actually on the carrying simplex? We may eliminate z since $z = 1 - x - y$ on the

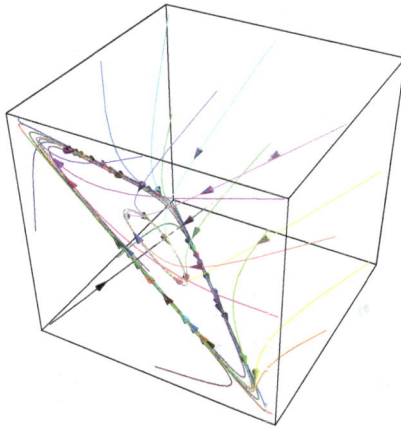

Fig. 4. Periodic orbits in a model of May and Leonard.[11] Note the carrying simplex is the usual simplex in $\mathbb{R}^3_{\geq 0}$ and it clearly attracts all orbits apart from the origin.

carrying simplex. This gives

$$\dot{x} = x(1 - x - \alpha y - \beta(1 - x - y)) = \frac{(\alpha - \beta)}{2}x(1 - x - 2y),$$

$$\dot{y} = y(1 - \beta x - y - \alpha(1 - x - y)) = \frac{-(\alpha - \beta)}{2}y(1 - 2x - y),$$

where $\alpha + \beta = 2$. Notice that div $(\dot{x}, \dot{y}) = 0$ and that we have a canonical Hamiltonian system with Hamiltonian function

$$H(x, y) = \frac{(\alpha - \beta)}{2}(1 - x - y)xy.$$

On the open triangle $T = \{(x, y) \in \mathbb{R}^2_{\geq 0} : 0 < x + y < 1\}$ we obtain closed contours, i.e. the solutions are periodic. (This is the projection of the dynamics on Σ onto the xy-plane.) Figure 4 shows the periodic orbits on the invariant plane $\Sigma = \{x \in \mathbb{R}^3_{\geq 0} : x_1 + x_2 + x_3 = 1\}$ as part of the three-dimensional phase portrait.

Problem 11.3. *Consider the planar competitive system*[20]

$$\dot{x} = x\left(1 - x - \frac{y}{2}\right),$$

$$\dot{y} = y(1 - 3x - y).$$

Show that this system has a carrying simplex Σ which is the graph of a quadratic function q that satisfies $q(0) = 1$ and $q(1) = 0$. Sketch the phase plane.

12.   Further Reading

The book by Takeuchi (Ref. 16) provides a comprehensive study of Lotka–Volterra equations. Hofbauer and Sigmund[10] contains much on Lotka–Volterra systems, plus also their close cousins the Replicator equations from evolutionary game theory. The cooperative and competitive Lotka–Volterra models discussed here are a small subset of monotone dynamical systems which are covered in Ref. 9 by Hirsch and Smith. See also the monograph on monotone dynamical systems by Smith.[15] In 2003 a new geometrical approach to the study of Lotka–Volterra systems was initiated by Zeeman and Zeeman.[19] The carrying simplex of competition models originated in a 1988 paper by Hirsch.[8] The geometry of carrying simplices has been studied by Tineo[17] and more recently by Baigent,[1] where in the latter paper links are made to the geometrical approach of Zeeman and Zeeman.[19]

13.   Sketch Solutions to Problems

Problem 2.13 Solution:

$$\dot{x} = x(1 - x - y), \tag{13.1}$$

$$\dot{y} = y(-1 - y + x). \tag{13.2}$$

From (13.1) we see that for any initial condition, since $y(t) \geq 0$, for t large enough $0 \leq x(t) \leq 1$. But then, from (13.2) eventually $\dot{y} < 0$ and so we must have $y(t) \to 0$ as $t \to \infty$. It is thus clear that all orbits are bounded and each $w((x_0, y_0))$ is non-empty. If $(p_1, p_2) \in w((x_0, y_0))$, then for some $t_k \to \infty$, $p_2 = \lim_{k \to \infty} y(t_k) = 0$. Thus $w((x_0, y_0)) \subset \{(s, 0) : s \in [0, 1]\}$. If $(P_1(t), 0)$ is the forward orbit with $(P_1(0), 0) = (p_1, 0) \in w((x_0, y_0))$ and $p_1 > 0$ then by invariance $(P_1(t), 0) \in w((x_0, y_0))$ for all $t \in \mathbb{R}$. But $P_1(t)$ satisfies $\dot{P}_1 = P_1(1 - P_1)$ and so $p_1 = \lim_{k \to \infty} P_1(t_k) = 1$. Hence for any $(x_0, y_0) \in \mathbb{R}^2_{\geq 0} \setminus \{(0,0)\}$ we have $\omega((x_0, y_0)) = \{(1,0)\}$, and $\omega((0,0)) = \{(0,0)\}$.

Problem 2.19 Solution:
The system $\dot{x} = x(1 - x)$ has invariant sets $\{0\}$, $\{1\}$, $[0, 1]$, $\{0, 1\}$, $\mathbb{R}_{\geq 0}$, $\mathbb{R}_{\geq 0} \setminus \{1\}$. The forward invariant sets are $[s_1, s_2]$ for any $0 \leq s_1 \leq 1 \leq s_2$. $B(1) = \mathbb{R}_{>0}$.

Problem 4.8 Solution:
No, the converse does not hold. Take the system $\dot{x} = x(1 - \frac{x}{2} - \frac{y}{2})$ and $\dot{y} = y(-1 + x + \frac{y}{8})$. Then there is a steady state $(\frac{6}{7}, \frac{8}{7})$ which globally attracts $\mathbb{R}^2_{>0}$, with $\det A = \frac{7}{16} > 0$ and $b = -\frac{1}{2} < 0$ but $f = \frac{1}{8} > 0$.

Problem 5.2 Solution:
Set $x = \log N, y = \log P$. Then $\dot{x} = a - be^y, \dot{y} = ce^x - d$. Set $H(x, y) = ay + dx - ce^x - be^y$. Then $\frac{\partial H}{\partial x} = d - ce^x = -\dot{y}$ and $\frac{\partial H}{\partial y} = a - be^y = \dot{x}$. Then $\dot{H} = \frac{\partial H}{\partial x}\dot{x} + \frac{\partial H}{\partial y}\dot{y} = 0$, so that H is a constant along orbits.

Problem 7.1 Solution:
$\{f, g\} = \nabla f \cdot A \nabla g$, where $A^T = -A$. Then using the summation convention

$$\{g, h\} = a_{ij} \frac{\partial g}{\partial x_i} \frac{\partial h}{\partial x_j}.$$

Thus

$$\{f, \{g, h\}\} = a_{ij} \frac{\partial f}{\partial x_i} \frac{\partial}{\partial x_j} \left(a_{lk} \frac{\partial g}{\partial x_l} \frac{\partial h}{\partial x_k} \right)$$

$$= a_{ij} a_{lk} \frac{\partial f}{\partial x_i} \left(\frac{\partial^2 g}{\partial x_l \partial x_j} \frac{\partial h}{\partial x_k} + \frac{\partial g}{\partial x_l} \frac{\partial^2 h}{\partial x_k \partial x_j} \right)$$

$$= a_{ij} a_{lk} \left(\frac{\partial f}{\partial x_i} \frac{\partial h}{\partial x_k} \frac{\partial^2 g}{\partial x_l \partial x_j} + \frac{\partial f}{\partial x_i} \frac{\partial g}{\partial x_l} \frac{\partial^2 h}{\partial x_k \partial x_j} \right).$$

By cycling terms

$$\{g, \{h, f\}\} = a_{ij} a_{lk} \left(\frac{\partial g}{\partial x_i} \frac{\partial f}{\partial x_k} \frac{\partial^2 h}{\partial x_l \partial x_j} + \frac{\partial g}{\partial x_i} \frac{\partial h}{\partial x_l} \frac{\partial^2 f}{\partial x_k \partial x_j} \right)$$

and

$$\{h,\{f,g\}\} = a_{ij}a_{lk}\left(\frac{\partial h}{\partial x_i}\frac{\partial g}{\partial x_k}\frac{\partial^2 f}{\partial x_l \partial x_j} + \frac{\partial h}{\partial x_i}\frac{\partial f}{\partial x_l}\frac{\partial^2 g}{\partial x_k \partial x_j}\right).$$

Consider second derivatives of h the sum $\{f,\{g,h\}\} + \{g,\{h,f\}\} + \{h,\{f,g\}\}$; this gives

$$a_{ij}a_{lk}\frac{\partial f}{\partial x_i}\frac{\partial g}{\partial x_l}\frac{\partial^2 h}{\partial x_k \partial x_j} + a_{ij}a_{lk}\frac{\partial g}{\partial x_i}\frac{\partial f}{\partial x_k}\frac{\partial^2 h}{\partial x_l \partial x_j}$$

$$= a_{ij}a_{lk}\frac{\partial f}{\partial x_i}\frac{\partial g}{\partial x_l}\frac{\partial^2 h}{\partial x_k \partial x_j} - a_{ij}a_{lk}\frac{\partial g}{\partial x_i}\frac{\partial f}{\partial x_l}\frac{\partial^2 h}{\partial x_k \partial x_j}$$

$$= a_{ij}a_{lk}\frac{\partial^2 h}{\partial x_k \partial x_j}\left(\frac{\partial g}{\partial x_i}\frac{\partial f}{\partial x_l} - \frac{\partial f}{\partial x_i}\frac{\partial g}{\partial x_l}\right).$$

This value of this last expression is not changed by interchanging labels i, l, whereas the bracketed term changes sign, and so the expression must be zero.

Problem 8.6 Solution:
The conditions are $\det A = bf - ec > 0$ and $dc > fa, ea > db$. Since $ec > 0$ we must have $bf > 0$, and hence either (i) $b, f > 0$ or (ii) $b, f < 0$. If $b, f > 0$ then $\frac{c}{f}d > a$ and $\frac{e}{b}a > d$. From these it is clear that both conditions cannot be satisfied unless $a, d > 0$. If $b, f > 0$ and $a, d > 0$ then the condition is $\frac{b}{e} < \frac{a}{d} < \frac{c}{f}$. (Of course, even if an interior steady state exists, the cooperative system $(e, c > 0)$ may have unbounded orbits.)

Problem 11.3 Solution:
An invariant curve connecting $(1,0)$ and $(0,1)$ is a solution $y:[0,1] \to \mathbb{R}_{\geq 0}$ of $y'(x) = \frac{y(1-3x-y)}{x(1-x-\frac{y}{2})}$ that satisfies the boundary conditions $y(0) = 1$ and $y(1) = 0$. Let the quadratic be $y(x) = 1+ax+bx^2$. Then this satisfies the boundary condition $y(0) = 1$. To satisfy $y(1) = 0$ we need $0 = 1 + a + b$ so that we need $a = -1 - b$, and y takes the form $y(x) = 1 - (1 + b)x + bx^2 = (1 - x)(1 - bx)$. For this function

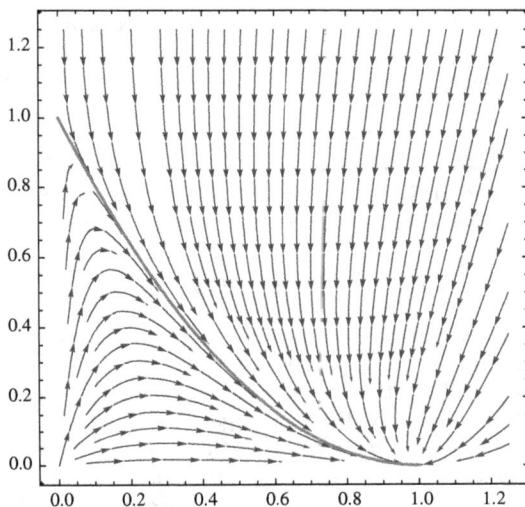

Fig. 5. Problem 11.3. There is a carrying simplex Γ that connects the two axial steady states $(1, 0)$ and $(0, 1)$. Γ is the graph of the function $y(x) = (1 - x)^2$ over $[0, 1]$ and attracts all points except the origin, which is a steady state.

$y'(x) = -1-b+2bx$, so that $y'(0) = -1-b$. On the other hand $y'(x) = \frac{y(1-3x-y)}{x(1-x-\frac{y}{2})}$, so by L'Hôpital's rule $y'(0) = \frac{y(0)(-y'(0)-3)+(1-y(0))y'(0)}{1-\frac{y(0)}{2}}$. Now set $y(0) = 1$ to obtain $y'(0) = -2$, which implies $b = 1$, and thus the curve is $y(x) = (1 - x)^2$. It is now simple to check that $y(x) = (1 - x)^2$ satisfies $y'(x) = \frac{y(1-3x-y)}{x(1-x-\frac{y}{2})}$ and so the graph of y is an invariant manifold. See Fig. 5 for the phase portrait.

References

1. S. Baigent, Geometry of carrying simplices of 3-species competitive Lotka–Volterra systems, *Nonlinearity* **26**(4), 1001–1029 (2013).
2. N. P. Bhatia and G. P. Szegö, *Stability Theory of Dynamical Systems*, Classics in Mathematics, Vol. 161. Springer Science & Business Media, (2002).
3. A. Berman and R. J. Plemmons. *Nonnegative Matrices in the Mathematical Sciences*, Classics in Applied Mathematics, Vol. 9. SIAM (1994).
4. C. Carmen, *Ordinary Differential Equations with Applications*. Texts in Applied Mathematics, Vol. 34. Springer-Verlag (2006).
5. P. Duarte, R. L. Fernandes and W. M. Olivia, Dynamics on the attractor of the Lotka–Volterra equations. *J. Differential Equations* **149**, 143–189 (1998).

6. B. S. Goh, Global stability in two species interactions, *J. Math. Biol.* **3**, 313–318 (1976).
7. B. S. Goh, Stability in models of mutualism, *Amer. Nat.* **113**(2), 261–275 (1979).
8. M. W. Hirsch, Systems of differential equations which are competitive or cooperative: III. Competing species, *Nonlinearity* **1**, 51–71 (1988).
9. M. W. Hirsch and H. L. Smith, Monotone dynamical systems. In *Handbook of Differential Equations, Ordinary Differential Equations*, Vol. 2. Elsevier, Amsterdam (2005).
10. J. Hofbauer and K. Sigmund, *Evolutionary Games and Population Dynamics*. Cambridge University Press (2002).
11. R. W. May and W. J. Leonard, Nonlinear aspects of competition between three species, *SIAM J. Appl. Math.* **29**, 243–253 (1975).
12. J. E. Marsden and T. Ratiu, *Introduction to Mechanics and Symmetry*, Texts in Applied Mathematics, Vol. 17. Springer-Verlag (1994).
13. L. Perko, *Differential Equations and Dynamical Systems*, Third Edition, Texts in Applied Mathematics, Vol. 7. Springer (2001).
14. S. Smale. On the differential equations of species in competition, *J. Math. Biol.* **3**, 5–7 (1976).
15. H. L. Smith, *Monotone Dynamical Systems*. American Mathematical Society, Providence, RI (1995).
16. Y. Takeuchi, *Global Dynamical Properties of Lotka–Volterra Systems*. World Scientific, Singapore (1996).
17. A. Tineo, On the convexity of the carrying simplex of planar Lotka–Volterra competitive systems, *Appl. Math. Comput.* **123**(1), 1–16 (2001).
18. V. Volterra. *Leçons sur la Théoerie Mathématique de la Lutte pour la Vie.* Gauthier-Villars et Cie., Paris (1931).
19. M. L. Zeeman and E. C. Zeeman, From local to global behavior in competitive Lotka–Volterra systems, *Trans. Amer. Math. Soc.* **355**, 713–734 (2003).
20. E. C. Zeeman, Classification of quadratic carrying simplices in two-dimensional competitive Lotka–Volterra systems, *Nonlinearity* **15**(6), 1993–2018 (2002).

Chapter 6

Applied Dynamical Systems

David Arrowsmith

School of Mathematical Sciences, Queen Mary University of London
London E1 4NS, UK
d.k.arrowsmith@qmul.ac.uk

In this chapter we consider the iteration of maps on the real line as dynamical systems in discrete time, and explore their representation in terms of shift maps on symbol sequences. Another key theme of the chapter is to explain links between dynamical systems and statistics through the dynamical phenomenon of intermittency and the auto-correlation decay behaviour of times series of binary data. The basic dynamical ingredients of orbits, fixed and periodic points, and dynamical equivalence of maps are introduced. We also show how symbolic dynamics can be used to reveal the intricate orbital structure of real and interval maps which when looked at numerically can have bewildering and chaotic behaviour. We introduce the concept of invariant densities for maps and their importance in describing natural distributions associated with typical orbits of dynamical systems. These insights are then used to discuss the concept of intermittency dynamics where orbits can alternate spasmodically between regular and chaotic behaviour. Finally, we show how intermittency can be used as a modelling tool to produce binary time series with various auto-correlation decay properties.

1. Dynamical Systems and Symbolic Coding

1.1. *Maps of the real line and the interval*

A *real map* is a function $f : \mathbb{R} \to \mathbb{R}$ where $\mathbb{R}$ is the set of real numbers. A map $f : \mathbb{I} \to \mathbb{I}$, where $\mathbb{I} = [0, 1] = \{x | 0 \leq x \leq 1\}$, is called an *interval map*. Thus repeated iteration of such maps given an initial point $x_0 \in \mathbb{I}$ (or $\mathbb{R}$) provides a sequence of real numbers $\{x_n\}$ in the interval $\mathbb{I}$ (or $\mathbb{R}$) where $x_n = f(x_{n-1})$, for $n \in \mathbb{Z}^+$.

A geometrical interpretation of the iteration of a map can be obtained by using the *orbit web*. The sequence of points $(x_n, x_{n+1}) \in \mathbb{R}^2$ gives the *dynamical iteration* of the orbit x_0 and the underlying map f is seen as a *dynamical system*. The dynamical iteration of f is obtained geometrically using the symmetry line $L : y = x$ to transfer the value $y = x_{n+1} = f(x_n)$, on $\mathrm{gr}(f)$, from the y-axis to the x-axis. This facilitates the next iteration of f to give the point $x_{n+2} = f(x_{n+1})$ and so on. Thus the "web" in $\mathbb{R} \times \mathbb{R}$ is obtained by alternately filling in vertical segments from (x_n, x_n) on L to (x_n, x_{n+1}) on $\mathrm{gr}(f)$ followed by horizontal segments from (x_n, x_{n+1}) to (x_{n+1}, x_{n+1}). This procedure moves back and forth between points on L and points on the graph of f and gradually unfolds the nature of an orbit as in Fig. 1.

1.2. *Behaviour of orbits*

A natural question to ask is what types of orbit can arise from functions such as those illustrated in Fig. 1? If the graphs $y = x$ and $y = f(x)$ intersect at a point $x = x^*$, then we have a *fixed* or *equilibrium* point of the iteration, i.e. $x^* = f(x^*)$ — there is no change in the iterated value. One can then ask about the behaviour of orbits which arise in the neighbourhood of the fixed point. Such orbits can move towards or away from the fixed point becoming an *attractor*, or

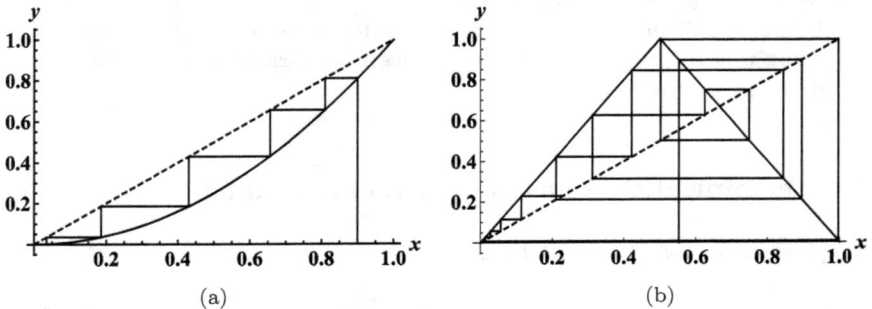

Fig. 1.   Orbit webs for various functions (a) $f(x) = x^2$; (b) $f(x) = 2x$, $x \in [0, 0.5)$ and $f(x) = 2 - 2x$ for $x \in [0.5, 1]$.

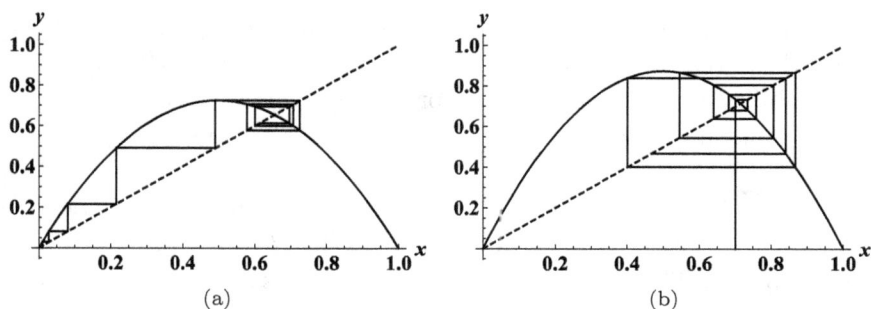

Fig. 2. Various orbital behaviours in the neighbourhood of fixed points: (a) $f(x) = 2.9x(1 - x)$: an unstable fixed point at $x = 0$, and a stable fixed point at $x \approx 0.7$, (b) $f(x) = 3.5x(1 - x)$: an unstable fixed point at $x \approx 0.7$.

repeller, respectively. Neutral fixed points can also arise where orbits from one side of x^* move towards the fixed point while they move away on the other side. Various configurations of fixed points can be seen in Fig. 2. They depend on the nature of the graph in the neighbourhood of its intersection with the 'fixed point' line. To use more precise language, we define a fixed point $x = x^*$ of f to be *stable* if given any neighbourhood U of the point x^*, there exists a second neighbourhood $V \subset U$ of x^* which has the property that for every orbit with initial point $x_0 \in U$ remains in V.

A stable point is *asymptotically* stable if it is the case that the neighbourhood V can be chosen such that $\lim_{n \to \infty} x_0 = x^*$, $x_0 \in V$. Such points are attractors.

If the map is differentiable then the stability behaviour at a fixed point can be partially categorised in terms of the derivative of f at the fixed point $x = x^*$. The linear Taylor expansion of the map f in the neighbourhood $x = x^*$ gives

$$f(x) \approx f(x^*) + f'(x^*)(x - x^*),\tag{1.1}$$

and so

$$|x_{n+1} - x^*| \approx |f'(x^*)||(x_n - x^*)|.\tag{1.2}$$

for $|x_n - x^*|$ sufficiently small.

Thus we have *asymptotic* stability for the fixed point x^* when $|f'(x^*)| < 1$ and *instability* when $|f'(x^*)| > 1$. For $|f'(x^*)| = 1$, the stability is not determined. For example, $f(x) = x - x^3$ has a stable point at $x^* = 0$, and $f(x) = x + x^2$ has an unstable point at $x^* = 0$.

Types of recurrence other than a fixed point can also occur. A point $x = x_0$ is a *periodic point* of f if $f^q(x_0) = x_0$ for some integer $q > 0$. The minimum such positive q is said to be the *period* of the orbit which consists of the finite set $x_0, x_1, \ldots, x_{q-1}$ which is then repeated infinitely since $x_q = x_0, x_{2q} = x_0$, etc. Note that $q = 1$ implies that the periodic orbit is simply a fixed point. The fixed point stability result extends to a q-periodic point x^* by considering its behaviour as a *fixed* point of the map f^q, $f^q(x_0) = x_0$. Thus, the period-q periodic orbit $\{x_0, \ldots, x_{q-1}\}$ of f is asymptotically stable if $|(f^q)'(x_0)| < 1$ and unstable, if $|(f^q)'(x_0)| > 1$. The apparent dependence of stability on just x_0 and no other points of the periodic orbit is illusory as the derivative at $x = x_0$ that determines stability is given by

$$(f^n)'(x_0) = \prod_{i=0}^{q-1} f'(x_i),\tag{1.3}$$

and so $(f^n)'(x_i)$ is the same for all points x_i of the periodic orbit. It is possible for a map f to become *eventually periodic*. An orbit with initial point x_0 can satisfy $f^q(x_k) = x_k$ for iterates $k > k_0$ only, for some fixed positive integer k_0. Note that if the map $f \colon \mathbb{I} \to \mathbb{I}$ has an inverse map $\bar{f} \colon \mathbb{I} \to \mathbb{I}$, then such an eventually periodic orbit of f after k iterates is immediately periodic, i.e. $f^q(x_0) = x_0$. This follows from applying the map $\bar{f}^k$ to both sides of the equation $f^q(x_k) = x_k$.

1.3.   *Equivalence of maps*

We can introduce an *equivalence* between maps which respects orbital behaviour. Let $\phi : \mathbb{R} \to \mathbb{R}$ be a homeomorphism, a map which is both bijective and bicontinuous. Suppose that we have two maps of the reals $f, g : \mathbb{R} \to \mathbb{R}$ such that the following *topological conjugacy* holds

$$g\phi(x) = \phi f(x),\tag{1.4}$$

for all $x \in \mathbb{R}$. We can deduce from the conjugacy that there is an equivalence of orbit structure for the maps f and g. For any initial point $x_0 \in \mathbb{R}$, the f-orbit $\{f^n(x_0)\}$ and the g-orbit $\{g^n(y_0)\}$ with $y_0 = \phi(x_0)$ are in one-to-one correspondence by the map ϕ because $\phi(f^n(x_0)) = g^n(y_0)$ from Eq. (1.4). Moreover, let x_0 be a periodic point of f of minimum period q, i.e. $f^q(x_0) = x_0$ and $f^k(x_0) \neq x_0$ for $0 < k < q$. Let $x_k = f^k(x_0)$, $k \in \mathbb{Z}^+$, then $\{x_0, x_1, \ldots x_{q-1}\}$ is the associated periodic orbit. Let $y_k = \phi(x_k)$. Then $\{y_0, y_1, \ldots y_{q-1}\}$ is a q-periodic orbit of the map g. Observe that the conjugacy equation (1.4) gives $g^k \phi(x) = \phi f^k(x)$, $k \in \mathbb{Z}^+$. Thus

$$g^q(y_0) = g^q(\phi(x_0)) = \phi f^q(x_0) = \phi(x_0) = y_0,$$

and

$$g^k(y_0) = g^k(\phi(x_0)) = \phi(f^k(x_0)) \neq \phi(x_0) = y_0$$

using ϕ is injective, and $f^k(x_0) \neq x_0$ for $1 \leq k < q$. It follows that y_0 is a period-q point of g. Thus ϕ maps periodic orbits of f to those of ϕ, and, in particular, fixed points of f to those of g. It can be similarly shown that aperiodic (i.e. non-periodic) orbits of f are mapped to the aperiodic orbits of g. *Eventually periodic orbits* of f and g are also in one-to-one correspondence. The bi-continuity of ϕ allows limiting structures of orbits to be associated. For example, the *limiting set* $L_f(x_0) = \lim_{n \to \infty} f^n(x_0)$, then $\phi(L_f(x_0)) = L_g(y_0)$, where $y_0 = f(x_0)$.

This equivalence extends to other classes of maps such as interval maps where the conjugating homeomorphism would be $\phi : \mathbb{I} \to \mathbb{I}$, for example, or some other subinterval of $\mathbb{R}$. The qualitative approach, which stems from a discussion of the properties invariant under equivalence by topological conjugacy, has led to the discovery of the key ingredients of dynamical chaos. An excellent introductory text for this is Ref. 5. First, we consider some simple maps for which the dynamical complexity is relatively easy to uncover and describe.

1.4. *Symbolic coding and shift maps*

We can see that a linear map $f(x) = \alpha x$ with $\alpha \in \mathbb{R}$ has trivial dynamical behaviour. The map has a fixed point at the origin $x = 0$

and all other points are aperiodic and orbits either move away from the origin, or towards it, depending on whether $|\alpha| > 1$ or $|\alpha| < 1$ respectively. For $\alpha = 1$, all points are fixed points, and for $\alpha = -1$, the origin is a fixed point and all other points have period 2. Orbital diversity is introduced only through nonlinearity.

The doubling map is the simplest nonlinear map of the interval $I = [0, 1)$ constructed by using *two* linear components, cf. Fig. 1(b). It can be expressed in the form

$$D(x) = \begin{cases} 2x, & 0 \le x < 1/2, \\ 2x - 1, & 1/2 \le x < 1, \end{cases} \tag{1.5}$$

or equivalently,

$$D(x) = 2x \ (\text{mod} \, 1). \tag{1.6}$$

If we see the map D as an iteration on $I = [0, 1)$ defined by

$$x_n = 2x_{n-1} \ (\text{mod} \, 1), \tag{1.7}$$

then it is not difficult to 'solve' this equation and deduce that the nth iterate $x_n = D^n(x_0)$ of $x_0 \in I$ is given by $x_n = 2^n x_0 \ (\text{mod} \, 1)$. Investigation of the iteration formula for x_n offers no more information or clarity than the defining equation (1.5) for D simply because it is not at all clear how the exponential doubling in the real numbers interacts with the process of reducing mod 1.

The temptation at this point is to resort to computation for more insight. Unfortunately, we have only a finite number, say m, of binary places stored in the machine for each real number we wish to represent in $[0, 1)$ and we lose information at the rate of 1-binary place per iteration since, as we shall see, the map effectively shifts the binary expansion of a point by one binary place and deletes the integer part. Thus if the computer fills out the number to m places after each iteration in a controlled way by adding 0 in the mth place, we have zero after m-iterations! If the mth place is filled randomly, then after m iterations we have a random number generator for binary m digit integers.

Nevertheless, interpreting the real numbers in binary form leads to a theoretical understanding of the dynamics of the map D. The sequence of real numbers $\{x_n\}$ obtained by iterating any interval map f can easily be turned into a binary sequence with the use of an *output* map $s : \mathbb{I} \to \{0,1\}$ defined by

$$s(x) = \begin{cases} 0, & 0 \le x < \bar{d}, \\ 1, & \bar{d} \le x \le 1, \end{cases} \tag{1.8}$$

where the discriminator $\bar{d} \in (0,1)$. Thus, given an initial $x_0 \in I$, we obtain the real sequence $\{x_n\}$ in I and deduce the output binary sequence $\{y_n\} \subseteq \{0,1\}^\infty$, where $y_n = s(x_n)$. This construction, which we call *symbolic coding*, allows us to investigate the complexity of the orbit structure of interval maps by the use of *binary time series*.

Any real number $x_0 \in I$ can be written in the *binary form*

$$x_0 = \sum_{n=1}^{\infty} b_n 2^{-n}, \tag{1.9}$$

where $b_n = 0$ or 1. Thus formally, we can represent each point of I as a sequence $\sigma = \{b_n\}_{n=1}^\infty$. Let $\mathcal{S}$ denote the set of all such binary sequences. Now we note that

$$D(x_0) = D\left(\sum_{n=1}^{\infty} b_n 2^{-n}\right) = 2\sum_{n=1}^{\infty} b_n 2^{-n} \pmod 1$$

$$= b_1 + \sum_{n=2}^{\infty} 2b_n 2^{-n} \pmod 1 = \sum_{n=1}^{\infty} b_{n+1} 2^{-n} \pmod 1, \quad (1.10)$$

since b_1 is an integer. Thus formally we have a conjugacy $\phi(\{b_n\}_1^\infty) = \sum_{n=1}^{\infty} b_n 2^{-n}$, such that

$$D\phi = \phi\alpha, \tag{1.11}$$

where

$$\alpha\{b_n\} = \{b_{n+1}\}, \tag{1.12}$$

for $n \in \mathbb{Z}^+$. The map α is called a *shift*. This straightforward observation allows us to examine the periodic structure of the map D. We note that the period-q periodic points of α in $\mathcal{S}$ arise from precisely those binary representations for x_0 that repeat after q-digits and no fewer. Thus period-1 points of α are given by the repeating expansions

$$\sigma_0 = \{00000\ldots\} = \{\bar{0}\} \quad \text{and} \quad \sigma_1 = \{11111\ldots\} = \{\bar{1}\}.$$

Period-2 points are given by

$$\sigma_2 = \{010101\ldots\} = \{\overline{01}\} \quad \text{and} \quad \sigma_3 = \{101010\ldots\} = \{\overline{10}\}.$$

$$(1.13)$$

Clearly the sequences σ_0 and σ_1 are the binary representations of $x = 0$ and $x = 1$ respectively and we note $D(0) = 0$, and the point $1 \notin I$. The sequence σ_2 gives the point $x = 1/4 + 1/16 + 1/64 + \cdots = 1/3$ and σ_3 gives the point $x = 2/3$. Note either by applying α to the two symbols σ_2 and σ_3, or D to the corresponding real values $1/3, 2/3$ we have a single period-2 orbit. Specifically,

$$\alpha(\sigma_2) = \sigma_3, \quad \alpha(\sigma_3) = \sigma_2 \qquad (1.14)$$

in $\mathcal{S}$ and, correspondingly,

$$D(1/3) = 2/3, \quad D(2/3) = 1/3, \qquad (1.15)$$

in $\mathbb{I}$. Geometrically, the orbit web for a periodic orbit such as this closes up after two iterations and is a closed loop. We can see immediately that periodic points of all orders can be constructed for α in this way and therefore there are corresponding periodic orbits for the map D.

Points which have orbits which are *eventually* period one or two can be obtained by delaying the introduction of the recurrences given in the above symbolic sequences, e.g. the orbit $\sigma = 110101\bar{0}$ is eventually a fixed point since $\alpha^n(\sigma) = \bar{0}$ for $n \geq 6$. A key observation on the type of orbits available from such a map comes from noting that if $\{b_n\}_{n=1}^{\infty}$ is the binary sequence arising from the binary representation of the point x then $x < 0.5$ if $b_1 = 0$ and $x \geq 0.5$ if $b_1 = 1$. Given that iteration of D corresponds to shifting the symbols of the binary sequence, we see that it provides, *at a glance*, the movement of the itinerary of the orbit through the regions '0' representing the interval $I_0 = [0, 0.5)$ and '1' representing $I_1 = [0.5, 1]$.

Remark. This statement is only strictly true if we are careful about the ambiguity of binary representation. Note that $1/2 = 0.1\bar{0} = 0.0\bar{1}\ldots$. In fact, every real number whose infinite binary expansion finishes in consecutive zeroes can also be expressed using an infinite sequence of ones. If we always choose zeroes for these ambiguous cases (the *dyadic numbers*) then the itinerary described above is accurate. So we see that these piecewise linear maps can give rise to all possible binary sequences and so this demonstrates the flexibility of the maps to create all experimental cases of '0–1' binary data streams.

By investigating the graph of the map D we can see that sub-intervals of $\mathbb{I}$ can be located which give rise to the various sequences of binary data. For the map D, we have the following intervals of one and two symbols:

$$[0, 0.5) - \text{'0'}; \qquad [0.5, 1] - \text{'1'};$$
$$[0, 0.25) - \text{'00'}; \quad [0.25, 0.5) - \text{'01'}; \qquad (1.16)$$
$$[0.5, 0.75) - \text{'10'}; \quad [0.75, 1.0] - \text{'11'}.$$

There are eight symbol sequences of length 3 and each sequence arises from a unique closed-open interval of length $1/8$ etc.. For example, "000" is the identifier for any initial point $x_0 \in [0, 1/8)$. Obviously, an infinite binary sequence arises from a unique real value as the defining interval lengths reduce to zero. Carrying this through all the finite sequences gives the symbolic encoding of the map D.

Variations on the doubling map D are the *tent* map $T : \mathbb{I} \to \mathbb{I}$:

$$T(x) = \begin{cases} 2x, & 0 \le x < \dfrac{1}{2}, \\ 2 - 2x, & 1/2 \le x \le 1, \end{cases} \qquad (1.17)$$

and the m-fold *sawtooth* map $S_m : I \to I$:

$$S_m(x) = mx \pmod 1, \qquad (1.18)$$

where $m \ge 2$ is a positive integer, and, again, $I = [0, 1)$. The map names reflect the shape of their graphs. Note that the doubling map D is also given by the special sawtooth map S_2. The tent map has a symbolic coding on 2-symbols, and can be shown to have similar periodic structure to the doubling map. The sawtooth requires

m-symbols, one for each of the intervals $[\frac{k}{m}, \frac{k+1}{m}]$, $k = 0, 1, \ldots, m-1$. We can then deduce that the map S_m has $m - 1$ fixed points, $\{k/m \mid 0 \le k < m\}$.

1.5. *Chaos in maps*

A crucial feature of the doubling map D on the interval I is the way in which neighbouring orbits move apart. More precisely, the map offers *sensitive dependence on initial conditions* (SDIC).[3] All of the above maps iterate to give *chaotic* behaviour:

(1) the map has SDIC;
(2) the map has dense orbits, that is, the orbit passes arbitrarily close to every point of the interval;
(3) the set of periodic points of D is dense in the interval.

Specifically, the SDIC condition for a chaotic map f requires that there exist some constant δ and points x' arbitrarily close to any given point x such that $\mathrm{dist}(f^n(x), f^n(x')) \ge \delta$ for some positive integer n. Note that given any two distinct points of x, x' of I, their orbits initially diverge exponentially in the sense that $\mathrm{dist}(D^n(x), D^n(x')) = 2^n \, \mathrm{dist}(x, x')$ subject, of course, to the global constraint here that no two points can ever be more than distance one apart on I. Thus, it is only *locally* true that orbits are diverging exponentially. Nevertheless, SDIC, as in Ref. 3, is certainly satisfied by the exponential divergence exhibited by maps such as D and T with, say, $\delta = \frac{1}{2}$.

All of the maps with two (or more) continuous segments can be shown to exhibit the flexible orbital behaviour of the chaotic doubling map. The doubling, tent and quadratic maps are similar symbolically, even though their graphs are obviously different. But all three interval maps stretch the interval twice over itself. An, apparently, similar type of map $g(x) = 3x \pmod 1$ on I is, in fact, different, as it requires *three*, rather than two, symbols to describe orbital itineraries with '0' = $[0, 1/3)$, '1' = $[1/3, 2/3)$, '2' = $[2/3, 1)$. The map S_3 differs from the doubling map in its periodic point structure. For example, S_3 has the following three period-2 periodic orbits:

$$\{\{\overline{01}\}, \{\overline{10}\}\}, \quad \{\{\overline{02}\}, \{\overline{20}\}\}, \quad \{\{\overline{12}\}, \{\overline{21}\}\}, \tag{1.19}$$

whereas D has just one period-2 periodic orbit, namely $\{\{\overline{01}\}, \{\overline{10}\}\}$. This observation shows that the maps S_3 and D are not conjugate as there is not a one-to-one correspondence of period-2 orbits. Similar simple considerations will show that S_m and S_n are not conjugate for $m \neq n$. This distinction can also be seen by S_m that have exactly $m - 1$ fixed points.

The symbolic coding allows us to check the remaining *Devaney* conditions for chaos (ii) and (iii).

The sensitive dependence on initial conditions shows up directly from the nature of the map D — the images of two points distance d apart are $2d$ apart when d is sufficiently small.

The existence of dense orbits follows from the construction of a symbol sequence containing all possible symbolic words. This can be done by listing in sequence all possible symbol sequences of all possible lengths.

The density of periodic orbits in I can be shown by observing that any real number $x \in [0, 1)$ with symbol sequence $\{b_n\}_{n=1}^{\infty}$ can be approximated increasingly well by symbolic sequences which are periodic with longer and longer periods.

2. Invariant Densities and Measures

Let us consider the output from two maps with similar graphs illustrated in Fig. 3

$$\text{(a)} \quad \bar{D}(x) = \begin{cases} 2x & \text{if } 0 \leq x < 0.5, \\ 2x - 1 & \text{if } 0.5 \leq x \leq 1, \end{cases} \tag{2.1}$$

and

$$\text{(b)} \quad f(x) = \begin{cases} x + 2x^2 & \text{if } 0 < x < 0.5, \\ 2x - 1 & \text{if } 0.5 \leq x \leq 1. \end{cases} \tag{2.2}$$

How can the different behaviour be identified from characteristics of these two maps? To answer this we have to consider how orbits distribute themselves both temporally and spatially and whether this can be characterised for almost all orbits, i.e. for a wide choice of

(a)

(b)

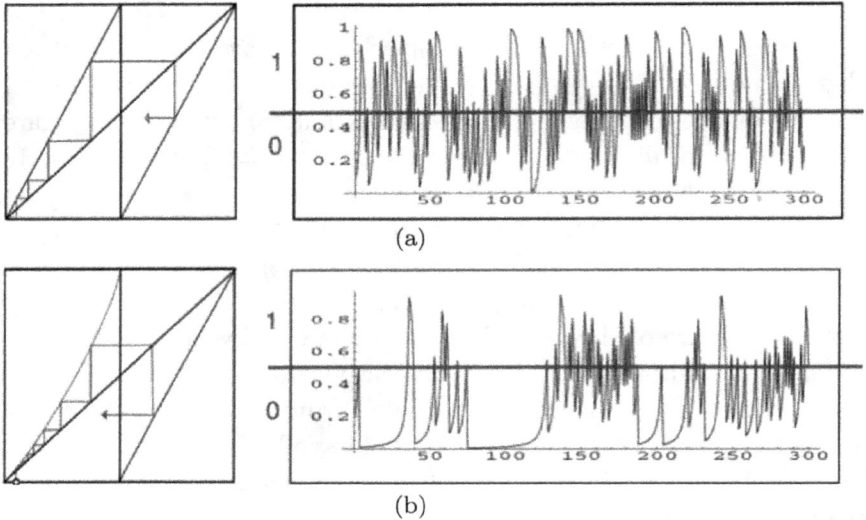

Fig. 3. Orbits for random initial conditions for maps (a) $\bar{D}$, and (b) f defined by Eq. (2.1) and Eq. (2.2), respectively. For each map we have the real orbital data which shows the time spent by the orbit in the regions $0 = [0, 0.5]$ and $1 = [0.5, 1]$. There are clear differences: the digital crossover between regions '0' and '1' is much more frequent, or *bursty*, for map (a) and the lengths of spells of consecutive '0's are much longer for map (b).

initial conditions. For this we need to develop the idea of *invariant densities* and the related concept of invariant measure for a map.

2.1. *Orbital densities and the Perron–Frobenius operator*

In Fig. 4, we consider the distributions of points generated by orbits of (a) the map $\bar{D}$, and (b) the quadratic map Q. The map $\bar{D}$ is iterated for $M = 10^6$ times and the map Q is iterated for $M = 10^5$ times. The unit interval is divided into N small equal length intervals where $N = 50$. We see an even distribution of visits to the intervals for the map $\bar{D}$ while the distribution Q shows typically orbits greater time close to $x = 0$ and $x = 1$. There is a 'smoothing' of the distribution as the number of iterations M is increased for both maps.

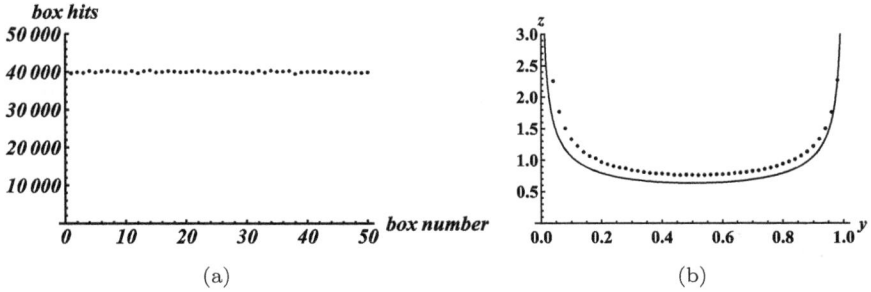

(a)   (b)

Fig. 4.   Distributions of orbital points into 50 partitioned subintervals of $\mathbb{I}$ for (a), the map $\bar{D}(x)$, and (b), the quadratic map, $Q(x) = 4x(1 - x)$, obtained from iterating the respective maps. These indicate the natural distributions that occur for almost all initial conditions. The graph of the invariant distribution, $z = \rho_Q(y) = 1/(\pi\sqrt{(y(1 - y))})$, of Q is shown superimposed to scale in (b).

It can be shown theoretically that the apparent smoothing of the distribution in Fig. 4(a) to a constant function actually occurs. Consider a probability distribution ρ_0 on the interval $\mathbb{I}$ which describes the distribution of a large ensemble of initial conditions for the map $\bar{D}$. Let the evolution of the ensemble ρ_0 after n iterations of f be given by the probability distribution ρ_n. The effect of one iteration on ρ_n is

$$\rho_{n+1}(x) = \int_0^1 \delta(x - f(z))\rho_n(z) \, dz, \tag{2.3}$$

where we use the *delta function* on $\mathbb{R}$ given by

$$\delta(x) = \begin{cases} \infty & \text{if } x = 0, \\ 0 & \text{otherwise,} \end{cases} \tag{2.4}$$

and

$$\int_{-\infty}^{\infty} \delta(x) = 1. \tag{2.5}$$

A justification for this result can best be seen by thinking of the distribution as arising from an orbit count. Let us consider the orbit $\{x_n\}_{n=0}^{\infty}$. Then the probability of orbital values taking the value y is

$$\rho_0(y) = \lim_{N \to \infty} \frac{1}{N} \sum_{i=1}^{N} \hat{\delta}(y - x_i), \qquad (2.6)$$

when we use the *discrete delta* function $\hat{\delta}$ which satisfies

$$\hat{\delta}(x) = \begin{cases} 1 & \text{if } x = 0, \\ 0 & \text{otherwise.} \end{cases} \qquad (2.7)$$

Consider $\hat{\delta}(y - f(x))\hat{\delta}(x - x_i)$ which takes the value 1 when both $y = f(x)$ and $x = x_i$, and zero otherwise. Thus

$$\hat{\delta}(y - f(x))\hat{\delta}(x - x_i) = \hat{\delta}(y - x_{i+1}). \qquad (2.8)$$

The distribution of points ρ_0 evolves to the distribution ρ_1 where (cf. 2.6)

$$\rho_1(y) = \lim_{N \to \infty} \frac{1}{N} \sum_{i=1}^{N} \hat{\delta}(y - x_{i+1}), \qquad (2.9)$$

since the set of points $\{x_i\}_{i=1}^{\infty}$ evolves to $\{x_{i+1}\}_{i=1}^{\infty}$. Using (2.8), we obtain

$$\rho_1(y) = \lim_{N \to \infty} \frac{1}{N} \sum_{i=1}^{N} \hat{\delta}(y - x_{i+1}) = \lim_{N \to \infty} \frac{1}{N} \sum_{i=1}^{N} \hat{\delta}(y - f(x))\hat{\delta}(x - x_i)$$

$$= \int \hat{\delta}(y - f(x)) \lim_{N \to \infty} \frac{1}{N} \sum_{i=1}^{N} \hat{\delta}(x - x_i) dx$$

$$= \int \hat{\delta}(y - f(x))\rho_0(x) dx,$$

which confirms Eq. (2.3) (cf. Fig. 5).

If the functional relation $\rho_0 \mapsto \rho_1$ is written as $\rho_1 = \mathcal{P}(\rho_0)$, it is known as the *Perron–Frobenius* equation and the operator $\mathcal{P}$ is known as the *Perron–Frobenius operator*. The equation can be written in various forms. For example, suppose that the graph f is

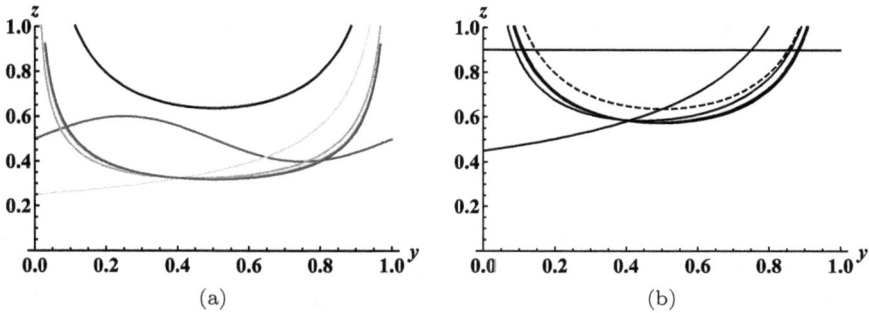

Fig. 5. The first few iterates of the Perron–Frobenius operator $\mathcal{P}_Q$ for the quadratic map $Q(x) = 4x(1-x)$: (a) the initial density is $\rho_0(y) = 0.5 + 0.1\sin(2\pi y)$ and (b) $\rho_0(y) \equiv 0.9$. Note the slowness of convergence of the iterates in (a) by comparison with (b) towards the natural invariant density $\rho_Q(y) = 1/\{\pi\sqrt{(y(1-y))}\}$ (dashed line).

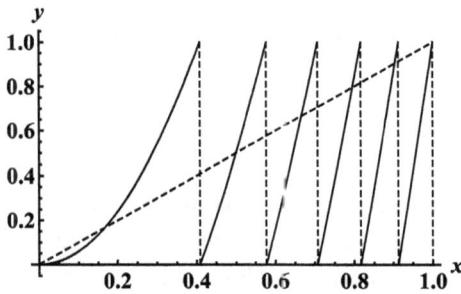

Fig. 6. The function $f(x) = 6x^2 \bmod 1$ on $[0,1)$ splits into six component homeomorphisms $f_1, f_2, \ldots, f_6$ on intervals $I_1, I_2, \ldots, I_6$ which are each surjective onto the interval $[0,1)$. Each component function $y = f_i(x)$ has an inverse $x = g_i(y)$.

composed of several continuous components

$$f_1, f_2, \ldots, f_n,$$

each of which has function inverses $g_1, g_2, \ldots, g_n$ on intervals $I_1, I_2, \ldots, I_n$ respectively (cf. Fig. 6). Then $f^{-1}(y) = \{g_1(y), \ldots, g_n(y)\}$ and we have

$$\rho_1(y) = \int_{x \in f^{-1}(y)} \rho_0(x)\,dx = \sum_{i=1}^{n} \rho_0(g_i(y))|g_i'(y)|, \qquad (2.10)$$

as when the functions g_i are differentiable, $g_i'(y)f_i'(x) \equiv 1$ with $y = f_i(x)$, and so

$$\rho_1(y) = \mathcal{P}(\rho_0)(y) = \sum_{i=1}^{n} \frac{\rho_0(x_i)}{|f'(x_i)|}, \qquad (2.11)$$

if $f^{-1}(y) = \{x_1, x_2, \ldots, x_n\}$, a finite set.

If the probability distribution ρ_0 does not change under iteration of Eq. (2.11), i.e. $\rho = \mathcal{P}(\rho)$, it is referred to as an *invariant* density and ρ satisfies

$$\rho(y) = \int_0^1 \delta(y - f(z))\rho(z)\, dz, \qquad (2.12)$$

and Eq. (2.12) is Perron–Frobenius identity for invariant measures. There can be many invariant densities. For example, let $\{x_0, x_1, \ldots, x_{q-1}\}$ be a period-q orbit of a map f. Then

$$\rho(x) = \begin{cases} 1/q & \text{if } x = x_i, \\ 0 & \text{otherwise} \end{cases} \qquad (2.13)$$

is an invariant probability density. However, this particular density is singular in the sense that it is **zero** almost everywhere in the interval. For the doubling map, orbital calculations can result in information on measures. The invariant measures for the doubling map split into three types. The discrete measures arise for (a) periodic and eventually periodic points, (b) typical, or *normal* irrational points,[4,5] and finally (c), atypical irrational points.

2.2. *Examples of invariant densities*

The doubling and tent maps: Let $y \in I$; then $D^{-1}(y) = \{y/2, (1+y)/2\}$ for the doubling map D and $D'(y/2) = D'((1+y)/2) = 2$. For the map T, $y \in \mathbb{I}$, $T^{-1}(y) = \{y/2, (2-y)/2\}$ and $|T'(y/2)| = |T'((2-y)/2)| = 2$. Then, we check Eq. (2.12)

$$\rho_T(y) = \frac{\rho_T(y/2)}{2} + \frac{\rho_T((1+y)/2)}{2} \qquad (2.14)$$

is satisfied by the constant map $\rho_T(x) \equiv 1$. The argument is the same for T and also can be easily extended to the sawtooth map S_m, for $m > 2$.

The quadratic map: The function $\rho_Q(y) = 1/(\pi\sqrt{(y(1-y))})$ satisfies the Perron–Frobenius equation (2.12) for the quadratic map Q.

Note that the preimages of the point y by the map Q are $x^* = (1 \pm \sqrt{(1-y)})/2$ and $f'(x^*) = 4 - 8x^* = \mp 4\sqrt{(1-y)}$. The Perron–Frobenius identity

$$\rho_Q(y) = \frac{1}{\pi}\frac{1}{\sqrt{y(1-y)}}$$

$$= \frac{\rho_Q((1 + \sqrt{(1-y)})/2)}{|-4\sqrt{(1-y)}|} + \frac{\rho_Q((1 - \sqrt{(1-y)})/2)}{|4\sqrt{(1-y)}|} \quad (2.15)$$

can be checked to confirm the invariance of the density ρ_Q.

Densities induced by conjugacy: An alternative approach to finding the invariant density for the map Q is to consider the conjugacy which relates it to the tent map T and transfer the invariant density from T to Q using the conjugacy.

Consider the map $\phi : \mathbb{I} \to \mathbb{I}$ where

$$\phi(x) = \sin^2\left(\frac{\pi x}{2}\right).$$

Observe that for $x \in [0, 0.5]$

$$Q\phi(x) = 4\sin^2\left(\frac{\pi x}{2}\right)\left(- \sin^2\left(\frac{\pi x}{2}\right)\right)$$

$$= \sin^2\left(2\frac{\pi x}{2}\right) = \phi(2x) = \phi(T(x)). \quad (2.16)$$

For the interval $x \in [0.5, 1]$,

$$Q\phi(x) = \sin^2\left(2\frac{\pi(1-x)}{2}\right) = \phi(2 - 2x) = \phi T(x). \quad (2.17)$$

Thus ϕ provides a conjugacy between the maps T and Q with $Q\phi(x) \equiv \phi T(x)$. The map $y = \phi(x) = \sin^2\left(\frac{\pi x}{2}\right)$ is a differentiable change of coordinates and the natural density $\rho_T(x) \equiv 1$ of the map T is transferred to the invariant density ρ_Q of the map Q via the conjugacy $\phi(x) = \sin^2(\frac{\pi x}{2})$. Observe

$$\frac{d\phi(x)}{dx} = \pi \sin\left(\frac{\pi x}{2}\right) \cos\left(\frac{\pi x}{2}\right) = \pi\sqrt{(y(1-y))}. \qquad (2.18)$$

Therefore

$$\frac{dy}{\pi\sqrt{y(1-y)}} = 1.dx \qquad (2.19)$$

and so the invariant densities $\rho_T(x) \equiv 1$ and $\rho_Q(y) = 1/(\pi\sqrt{(y(1-y))})$ are equivalent.

Normal and non-normal irrational points: For the doubling map the *natural* invariant measure arises from the orbital distributions associated with normal initial points. The normal points form a set of measure one in $\mathbb{I}$. It should be noted that periodic points which form the singular measures are also dense in $\mathbb{I}$ for the map D but more importantly, they have zero measure. Essentially, all finite patterns of digits are equally likely to occur in its binary expansion.[5] Thus normal numbers must be irrational as rational numbers have cyclic binary expansions and therefore a restricted set of digital patterns. The points of the orbit of such a normal number would therefore spread evenly over the interval and the *natural* invariant measure ρ for the doubling map D is uniform, i.e. $\rho(x) \equiv 1$ which satisfies invariance of the Perron–Frobenius equation.

It is easy to construct non-periodic orbits which do not have uniform distribution on the interval $\mathbb{I}$. For example, the orbit of the point x_0 represented symbolically by

$$x_0 = 01001000100001000001\ldots$$

clearly spends increasingly most of its time in the '0' region when iterated by D. In fact, asymptotically, the proportion of its time spent in the '0' region tends to 1, and is an example of a non-normal irrational number.

2.3.   *Invariant measures*

Given an invariant density ρ for $f : X \to X$, where $X \subseteq \mathbb{R}$, we can consider an associated *measure* μ where

$$\mu(S) = \int_S d\rho = \int_S \rho(x)dx \qquad (2.20)$$

on subsets S of X.

This measure also has invariance properties associated with the map f. Consider

$$\mu(f^{-1}(S)) = \int_{f^{-1}(S)} \rho(x)dx. \qquad (2.21)$$

The relation $y = f(x)$ implies $y \in S \iff x \in f^{-1}(S)$. The invariant density ρ implies $\rho(y)dy = \rho(x)dx$ and so we deduce

$$\mu(f^{-1}(S)) = \int_{f^{-1}(S)} \rho(x)dx = \int_S \rho(y)dy \qquad (2.22)$$

and the measure invariance for f can be written as

$$\mu(f^{-1}(S)) = \mu(S). \qquad (2.23)$$

Thus the measure is preserved for inverse images. This is true regardless of whether or not f has a map inverse. For example with the doubling map D, or the tent map T, the inverse image of an interval of length d is the union of two intervals each of length $d/2$. More precisely, if $S \subseteq \mathbb{I}$ is measurable, then $T^{-1}(S) = S^{(1)} \cup S^{(2)}$ where $S^{(1)} = \{x/2 | x \in S\}$ and $S^{(2)} = \{(2 - x)/2 | x \in S\}$. Given the uniform measure on $\mathbb{I}$, we have $\mu(S^{(1)}) = \mu(S^{(2)}) = \mu(S)/2$, and $\mu(S^{(1)}) + \mu(S^{(2)}) = \mu(S)$ which shows the measure is preserved.

The property does not extend to forward images, whose measure can, for example, double under iteration by T.

3.   Intermittency

We have seen how dynamical systems on the real line can produce different types of orbital distribution with maps such as that given by Eq. (2.2) and illustrated in Fig. 3. This is an example of *intermittency* in maps where the incremental steps can be small and prolonged in an orbit. We are able to relate the degree of intermittency in such maps to the statistics of their digital output sequences.

More specifically, the *auto-correlation* of a binary time series $\{x_t\}_{t=0}^{\infty}$ describes the correlation between sequence values at different times as a function of the lag $k = t_1 - t_2$ between two times t_1, t_2. Typically the correlation (cf. Eq. (3.11)) decays with increasing k and the decay can takes several forms. *Power-law* decay of the auto-correlation with increasing k is said to have *memory* or *long-range dependence* (LRD), whereas the relatively fast *exponential decay* is seen to have *no memory* or *short-range dependence* (SRD).[2] We now develop the relationship between dynamical intermittency and power law auto-correlation decay.

Intermittency, investigated by Pomeau and Manneville,[6] is achieved for real or interval maps by having segments of gr(f) given by $y = f(x)$ arbitrarily close, or even tangent, to the fixed point line $y = x$. In Fig. 7 we see two types of behaviour from this viewpoint. In (a), a graph which is close to the line $y = x$ provides a region of small iterative steps and a string of constant digital output. If the graph is relatively remote from the line $y = x$ then larger iterative steps occur. By comparison in (b), the graph of the map f meets the line $y = x$ in a fixed point $x = x^*$ of f. In this case, infinitely many

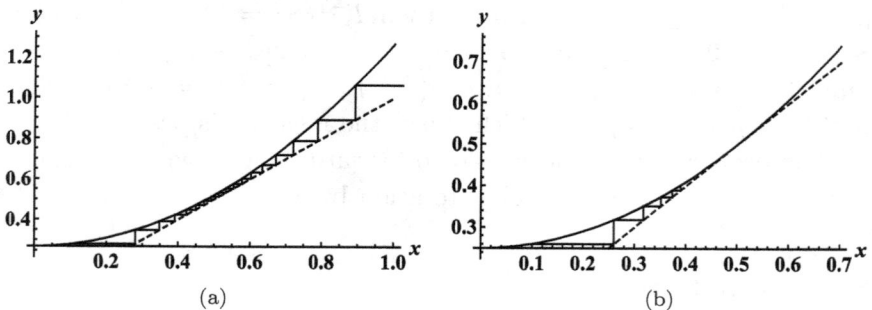

Fig. 7. Orbital behaviour of functions which increment slowly under iteration (a) $f(x) = x^2 + 0.27$, and (b) $f(x) = x^2 + 0.25$.

iterations from an initial point x_0 accumulate at x^* and the iteration takes increasingly small steps. Note that such a scenario could produce an infinity of either consecutive zeroes or ones depending on the location of the point x^*. These observations are key to producing output with 'long memory' and output with 'short memory'. To show how these geometrical cases provide different statistical behaviour in the binary output, we will consider a case of *intermittency* for a map at the origin.

Let $f : \mathbb{I} \to \mathbb{I}$ be defined by

$$f(x) = \begin{cases} x + (1-d)\left(\dfrac{x}{d}\right)^m, & 0 \le x < d, \\[2mm] \dfrac{x-d}{1-d}, & d \le x < 1. \end{cases} \qquad (3.1)$$

and the output $s : \mathbb{I} \to \{0, 1\}$ with

$$s(x) = \begin{cases} 0, & 0 \le x < \bar{d}, \\ 1, & \bar{d} \le x \le 1. \end{cases} \qquad (3.2)$$

Note that it is not necessary to have $\bar{d} = d$. We choose $m > 1$ to ensure an intermittency effect. We are interested in the lengths of consecutive sequences of '0's in the binary output. If we consider the simpler piecewise linear case of $g : \mathbb{I} \to \mathbb{I}$ with

$$g(x) = \begin{cases} \dfrac{x}{d}, & 0 \le x < d, \\[2mm] \dfrac{x-d}{1-d}, & d \le x \le 1, \end{cases} \qquad (3.3)$$

the slope of the first branch of g is $1/d$ and we can calculate immediately that a "zero" sequence will be of length k if the initial point x_0 lies in the interval $[\bar{d}d^k, \bar{d}d^{k-1}]$. Given that the natural invariant density of the map g is $\rho(x) = 1$, we have that the probability of a string of consecutive zeroes of *least* length k, given that x_0 is in $[0, \bar{d})$, is $P(k_{(\ge)}) \sim d^{k-1}$. Therefore, the probability of obtaining increasingly long strings of zeroes decays exponentially in the piecewise linear case since $0 < d < 1$ (cf. Fig. 8).

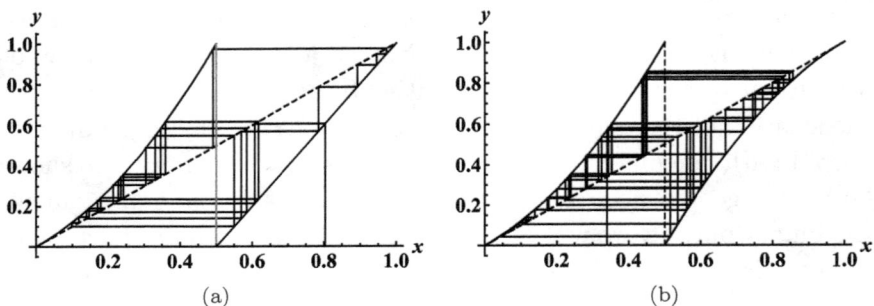

Fig. 8.   Orbits of piecewise continuous maps: (a) $f : \mathbb{I} \to \mathbb{I}$ defined in Eq. (3.1), with $d = 0.5$ and $m = 2$, exhibiting intermittency at $x = 0$; (b) $f : \mathbb{I} \to \mathbb{I}$ defined in Eq. (3.14), with $d = 0.5$, $m_1 = 2$ and $m_2 = 3$, exhibiting double intermittency at both $x = 0$ and $x = 1$.

3.1.   *Closed form intermittency*

We now return to the intermittent case. The simplest map exhibiting intermittency was given by $f : [0, d] \to [0, 1]$ where $f(x) = x + (1 - d)(\frac{x}{d})^m$. However, somewhat surprisingly, it is much easier to work with the map

$$\hat{f}(x) = \frac{x}{(1 - \hat{d}x^{m-1})^{\frac{1}{m-1}}}.$$

First of all, the map $\hat{f}$ belongs to the same family as f since an expansion gives

$$\hat{f}(x) = x + \frac{\hat{d}}{(m - 1)}x^m + O(x^{m+1}),$$

which has the same degree for its leading intermittency term as f. However, the map $\hat{f}$ has the striking property of its formula being algebraically closed under dynamical iteration. Secondly, it can be shown that

$$\hat{f}^k(x) = \frac{x}{(1 - k\hat{d}x^{m-1})^{\frac{1}{m-1}}}, \tag{3.4}$$

for $k = 0, 1, 2, \ldots$. Assuming Eq. (3.4), it can be checked directly that

$$\hat{f}^{k+1}(x) = \hat{f}(\hat{f}^k(x)) = \frac{x}{(1 - (k+1)\hat{d}x^{m-1})^{\frac{1}{m-1}}}. \tag{3.5}$$

The general formula (3.4) for k, a positive integer, then follows by induction noting that Eq. (3.4) gives $\hat{f}^0(x) = x$ as required. The formula can be extended further to rational and irrational values of k given the monotonic form of f with respect to both x and k.

Forcing the coefficients of x^m for f and $\hat{f}$ to be the same is not the appropriate identification for the two maps as $\hat{f}$ has infinite asymptotic behaviour when $\hat{d}x^{m-1} = 1$. The best approximation is obtained by requiring $\hat{f}(d) = 1$, which implies $\hat{d} = (1 - d^{m-1})/d^{m-1}$.

3.2. Power law and exponential escape

Suppose that we are now interested in the behaviour of orbital escape from the region $[0, \bar{d}]$ by iteration of the map $\hat{f}$. Let the point $x = x(k)$ 'escape to infinity' in exactly k iterations, i.e. $\bar{d} = \hat{f}^k(x(k))$. Solving for $x(k)$ we have

$$x(k) = \frac{\bar{d}}{(1 + k\hat{d}\bar{d}^{m-1})^{\frac{1}{m-1}}}.$$

As $k \to \infty$, we have

$$x(k) \sim K(\hat{d}, \bar{a})k^{\frac{-1}{m-1}},$$

for some function $K(\hat{d}, \bar{d})$. By solving the above equation for k, we see that the number of iterations for the orbit initially at x to escape is

$$k(x) = \frac{1}{\hat{d}} \left(\frac{1}{x^{(m-1)}} - \frac{1}{d^{(m-1)}} \right) \tag{3.6}$$

and so we obtain

$$k(x) \sim \frac{1}{\hat{d}}x^{-(m-1)},$$

as $x \to 0$. Strictly, we should take $\lceil k \rceil$ to obtain the integer solution for the number of iterates to escape since $k \in \mathbb{R}$. If $\hat{f}$ also has intermittency at $x = 1$, a similar integer function can be found which gives the number of consecutive ones for an initial condition $x_0 > d$. These equations are not available for the simpler functional form f (cf. Fig. 9).

We can calculate the probability of 'escape' from an intermittency region. Specifically, we consider the probability of a sequence

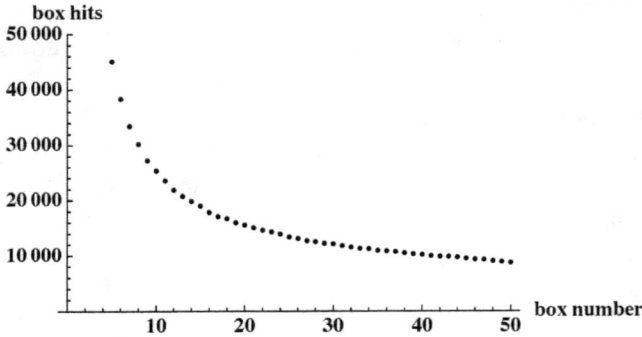

Fig. 9.   The distribution of an orbit of length $2 \cdot 10^6$ for the map given by Eq. (3.1), with $d = 0.5$ and $m = 2$, when $\mathbb{I}$ is partitioned into 50 equal subintervals. The introduction of the intermittency term at $x = 0$ to the piece-wise linear map $\bar{D}$ has the effect of changing its constant probability density function into a power law decay at the origin.

of k-consecutive zeroes for the output s of an intermittency map. We will consider the map $\hat{f}$ defined in Eq. (3.5), for $0 \le x < d$, and will assume a random injection to the region from $x \ge d$.[7]

If the orbit re-enters the interval $[0, d]$ at the point $\bar{x}$, then that determines the sequence length $l(\bar{x})$ of zeroes using Eq. (3.6). Furthermore, if $\hat{P}(\bar{x})$ is the probability density for least length l zero sequences, $l = k(\bar{x})$, then

$$\hat{P}(\bar{x}) = P(l(\bar{x}))\frac{dl}{d\bar{x}}, \qquad (3.7)$$

and if the re-entry density $\hat{P}(\bar{x})$ is assumed to be uniform, then

$$P(l) \sim \left|\frac{d\bar{x}}{dl}\right| = K''(\hat{d})l^{-\frac{m}{m-1}}, \qquad (3.8)$$

for some constant K'', which has a power law decay exponent of $m/(m-1)$.

The comparison of the intermittency and linear maps shows up in the nature of escape for the linear(exponential) case. We have seen that the escape in the former case depends on a power law relation. In the linear case (assume $f(x) = \alpha x$, with $\alpha > 1$), we are solving the corresponding escape equation $f^k(x) = \bar{d}$, i.e. $\alpha^k x = \bar{d}$, so we obtain $l = k(x) = \ln(\bar{d}/x)/\ln(\alpha)$.

With uniform injection of points into the interval, we have $P(l)$, the probability density for least length l zero sequences, $P(l(x)) \sim -\ln(x)/\ln(\alpha)$ and so with a change of variables

$$P(l_{(\geq)}) \sim \left| \frac{d\bar{x}}{dl} \right| = \frac{1}{\ln(\alpha)} e^{-l}. \tag{3.9}$$

We see the power law in Eq. (3.8) compares with an exponential law in Eq. (3.9).

3.3. *Auto-correlation for maps*

The *average* $E(G)$ of a real function $G : \mathbb{R} \to \mathbb{R}$ with respect to a real sequence $\{X_t\}_{i=1}^{\infty}$ is

$$E(G) = \lim_{N \to \infty} \frac{1}{N} \sum_{i=1}^{N} G(X_t) = \lim_{N \to \infty} \frac{1}{N} \sum_{t=1}^{N} \int G(X) \delta(X - X_t) \, dx$$

$$= \int G(X) \lim_{N \to \infty} \frac{1}{N} \delta(X - X_t) \, dX = \int G(X) \rho(X) dX. \tag{3.10}$$

We have already seen in Fig. 3 that the movement between strings of the output values '0' and '1' is rapid in the first trace and much slower in the second trace. The intermittency in one of the maps produces increased sojourn times for the two states. The longer sojourn times are said to introduce *memory* into the output which is reflected in higher correlation between the output binary sequence and the same sequence with a time-lag k. The nature of the decay of the auto-correlation vector is the way in which memory is measured.

Let X_t be a scalar time series of the binary values $\{0, 1\}$ for $t = 0, 1, 2, \ldots$, and suppose the series is stationary. We define the auto-correlation $\gamma(k)$ of lag k by

$$\gamma(k) = \frac{E(X_t X_{t+k}) - E(X_t) E(X_{t+k})}{\sqrt{(\text{Var}(X_t) \text{Var}(X_{t+k}))}}. \tag{3.11}$$

Stationarity of the sequence $\{X_t\}$ gives $\mu = E(X_t) = E(X_{t+k})$. Note also that because the X_t values are binary, $X_t^2 \equiv X_t$ and so

$E(X_t^2) = E(X_t) = \mu$ and $\text{Var}(X_t) = E(X_t^2) - E(X_t)^2 = \mu(1 - \mu)$. Therefore, the auto-correlation can be re-written as

$$\gamma(k) = \frac{E(X_t X_{t+k}) - \mu^2}{\mu(1 - \mu)}. \tag{3.12}$$

Given $0 \leq X_t X_{t+k} \leq X_t$, it follows that $\gamma(k) \leq 1$. In general, we expect that the correlation coefficient $\gamma(k)$ decays to zero as $k \to \infty$. If fact, if there were no correlation, i.e. the values X_t were *independent* of each other, then $E(X_t X_{t+k}) = E(X_t)E(X_{t+k}) = \mu^2$ and therefore $\gamma(k) \equiv 0$.

Two special types of decay are:

(a) power-law, where $\gamma(k) \sim ck^{-\beta}$ for some constants c and $\beta > 0$;
(b) exponential decay, where $\gamma(k) \sim c\alpha^{-k}$, for some constants c and $\alpha > 0$.

Let us consider the correlation behaviour of the doubling map D. Given a current state is '0', then $0 \leq x < 0.5$. The probability of the transfer '0 $\to$ 0' is 0.5, since it requires $0 \leq x < 0.25$, and similarly for '1 $\to$ 1'. Thus we can calculate exactly

$$\text{Pr}('0 \to 0' \text{ or } '1 \to 1') = 0.25 + 0.25 = 0.5$$

within one iteration. For two, or more, iterations,

$$\text{Pr}('0 \to 0' \text{ or } '1 \to 1') = 4 * 0.125 = 0.5,$$

and hence $E(X_t X_{t+k}) = 0.25$. The uniform density in the interval $\mathbb{I}$ gives $E(X) = 0.5$. It follows that $\text{Var}(X) = E(X^2) - E(X)^2 = E(X) - E(X)^2 = 0.25 - 0.5^2 = 0.25$. Therefore

$$\gamma(k) = \begin{cases} 1, & k = 0, \\ (0.25 - 0.5^2)/0.25 = 0, & k > 0. \end{cases} \tag{3.13}$$

The power-law behaviour extends to maps with competing intermittency behaviour. The map

$$f(x) = \begin{cases} x + (1 - d)\left(\dfrac{x}{d}\right)^{m_1}, & 0 \leq x < d, \\ x - d\left(\dfrac{1 - x}{1 - d}\right)^{m_2}, & d \leq x < 1, \end{cases} \tag{3.14}$$

with the extra condition that whenever f iterates across the line $x = \bar{d}$, the formula is replaced by a random uniform injection into the other interval. Note the competing intermittency provided by the parameters m_1 and m_2. The auto-correlation behaviour of f in Eq. (3.14) is given by

$$\gamma(k) \sim K'k^{-c}, \tag{3.15}$$

where $c = (2-m)/(m-1)$, $m = \text{Max}\{m_1, m_2\}$, and K' is a constant. Note that the stronger intermittency prevails in the autocorrelation as each side of the interval $\mathbb{I}$ is sampled with the iteration of f.

The relationship between intermittency and the auto-correlation decay exponent is now explicit and can be seen to be monotone.

Allowing for change from differentiable to piecewise linear maps, with the change from the intermittency parameter m in the smooth map case to the parameter α in the piecewise linear map, we have a similar result. Matching the asymptotics of the piecewise differential and piece-wise linear models requires the transformation $m = (\alpha + 1)/\alpha$.

If we consider a piecewise linear map constructed from two sequences of vertices in $\mathbb{I}^2$, $z_i = (i+1)^{-\alpha}, i^{-\alpha})$ at $x = 0$, and $w_i = (1 - (i+1)^{-\beta}, 1 - i^{-\beta})$ at $x = 1$, for $i = 1, 2, \ldots$, and $\alpha, \beta > 1$, then the two intermittencies compete and it can be proven that

$$\gamma(k) \sim Kk^{-\bar{c}},$$

where $\bar{c} = \text{Min}\{\alpha, \beta\} - 1$, K constant.[1,8] Thus the correlation for the composite map is determined by the heaviest tail (largest decay) in the correlation decay arising from the two competing intermittencies. The property was used in Setti *et al.*[8] to construct a binary generator of packet traffic in a network which simulated different levels of LRD and SRD packet traffic.

4. Exercises

1. Sketch functions $f : \mathbb{R} \to \mathbb{R}$ with exactly: (a) three fixed points — two unstable and one stable; (b) two fixed points — both unstable; (c) two fixed points — neither stable nor unstable.

2. Consider the function $f : \mathbb{I} \to \mathbb{I}$ given by $f(x) = (1+\sin(2\pi x))/2$. Find the fixed points and their stabilities.

3. For the map $\mathcal{D}(x) = 2x \pmod 1$, $x \in I = [0,1)$:

   (a) Find the number of points of period-1 , -2 and, general period-p;

   (b) Construct an orbit whose orbital density is the delta function with singularity at $x = 0.5$;

   (c) Construct an eventually periodic orbit of period-2 which stays within 2^{-5} of the origin the first 10 iterations, but is eventually period-2.

4. Consider $f_\mu : \mathbb{I} \to \mathbb{I}$, where $f_\mu(x) = \mu x(1 - x)$. Iterate f_μ, with $\mu = 1$, for a choice of initial points $0 < x_0 < 1$. Do all the orbits approach zero? Plot the graphs of f_1 and the identity function $I(x) = x$. Are there any fixed-points other than $x_0 = 0$? Repeat for f_μ with $\mu = 2$ and $\mu = 3$.

5. Show that $f_\mu(x) = \mu x(1 - x)$ with $\mu > 1$ has a unique fixed point $x_0(\mu) > 0$. Find the value of $\mu = \mu_0$ for which $x_0(\mu_0)$ changes from being stable for $\mu < \mu_0$ to unstable for $\mu > \mu_0$. Show that as μ increases through μ_0, a period-2 orbit is created.

6. Let $f : \mathbb{I} \to \mathbb{I}$ be a differentiable map with inverse $g : \mathbb{I} \to \mathbb{I}$. Furthermore, suppose that $x_0 \in \mathbb{I}$ satisfies $f^N(x_0) = x_0$ for some integer $N > 0$. Show that if $x_r = f^r(x_0)$, then $(f^N)'(x_0) = (f^N)'(x_r)$, for any $r \in \mathbb{N}$. Why does this make the stability of a periodic orbit $\{x_0, x_1, \ldots, x_N\}$ well-defined? Hence, or otherwise show that if f has a stable periodic orbit, then g has an unstable periodic orbit.

7. Consider the maps of the interval $[-1,1]$ given by (a) $f(x) = -1 + 2x^2$, and (b) $g(x) = -3x + 4x^3$. Show that they have a symbolic representation by using formulae for $\cos(n\theta)$ for $n = 2, 3$. What is the generalisation?

8. Let map $f : \mathbb{I} \to \mathbb{I}$ have a constant invariant density. Prove that if $N(x, (a, b), n)$ is the number of points among the first n points of the sequence $\{x, f(x), f^2(x), \ldots\}$ lying in the interval $(a, b) \subset \mathbb{I}$, then

$$\lim_{n \to \infty} \frac{1}{n} N(x, (a, b), n) = b - a. \tag{4.1}$$

9. Consider the *Newton–Raphson iteration* for $f : \mathbb{R} \to \mathbb{R}$ given by $x_{n+1} = f(x_n) = x_n - g(x_n)/g'(x_n)$ where $g : \mathbb{R} \to \mathbb{R}$ is a differentiable function. Show that a fixed point of the iteration corresponds to a zero of the function g. Prove that if g has a zero at $x = x^*$, with an initial approximation x_0 then a better approximation is given by $x_1 = f(x_0)$ such that $|x_1 - x^*| = K_n |x_0 - x^*|^2$ where $K_n = \frac{1}{2}|g''(\xi_n)/g'(\xi_n)|$. *Hint*: Use the Lagrange form of the remainder for Taylor series for g.

10. Show that the Newton–Raphson map for finding the square root of the integer n can be written in the form $f(x) = (x^2 + n)/(2x)$. With $n = 5$ and $x_0 = 1$, find the first three iterates.

5. Hints to Exercises

1. Consider functions on $\mathbb{R}$ such as (a) $f(x) = x^3$; (b) $f(x) = 1 - 2|x|$; (c) $f(x) = x + (x^2 - 1)^2$.

2. Note for $x \in [0, 0.5]$, $f(x) \geq 0.5 > x$, whereas for $x \in [0, 0.5]$, $f(x) \leq 0.5 < x$. The only fixed point is given for $x_0 = 1/2$ for which $f'(x_0) = -\pi$ and so the fixed point is unstable.

3. (a) • Symbolically, the fixed point at $0.000\ldots$ is written $\overline{0}$; period-2 points are symbolically $\overline{01}$ and $\overline{10}$ giving one periodic orbit of period-2.
   • Period-3 points are $\overline{100}$, $\overline{010}$, $\overline{001}$, and $\overline{110}$, $\overline{011}$, $\overline{101}$, giving two periodic two period orbits of period-3. Note $\overline{000}$ and $\overline{111}$ are period-1 but not period-3 orbits.

   (b) In general, proof recursively that for $p > 1$,

   $$N(p) = \frac{1}{p}\left(2^p - \sum_{k \text{ divides } p} k \times N(k)\right)$$

   using the idea that repetitive periods of lower order are not to be counted. For example $\overline{0101}$ is not a period-4 orbit, but a period-2 orbit.

   (c) Consider the orbit of $x_0 = 0.00000000\overline{10}$ which is eventually the periodic orbit $\{1/3, 2/3\}$.

4. All orbits $\{x_n\}$ of f_μ approach zero as $n \to \infty$ for $\mu \leq 1$. Note $0 < x_{n+1} = x_n(1 - x_n) < x_n < x_{n-1} < \cdots < x_0 < 1$. Therefore $\{x_n\}$ is a monotonic sequence bounded below by 0 and has a limit, say x^*. Note $\lim_{n\to\infty} x_n = x^* \implies x^* = x^*(1 - x^*) \implies x^* = 0$. For $\mu = 2$ the fixed points are given by $x = 0$ and $x = 1/2$. For $\mu = 3$, the fixed points are given by $x = 0$ and $x = 2/3$.

5. Note that $(f^N)'(x_0) = \prod_{i=0}^{N} f'(x_i)$ where $x_i = f^i(x_0)$ and the elements $f'(x_i)$ can be permuted to get

$$(f^N)'(x_0) = \prod_{i=0}^{N} f'(x_i) = \prod_{i=r}^{r+N} f'(x_i) = (f^N)'(x_r).$$

Therefore, the stability property is shared by all points of the periodic orbit. If $g = f^{-1}$, then $g'(f(x)) = 1/f'(x)$, and $(f^N)'(x_0) = \prod_{i=0}^{N} f'(x_i) = \prod_{i=0}^{N} 1/g'(x_{i+1}) = 1/(g^N)'(x_0)$. The derivatives are reciprocal for f and $g = f^{-1}$ interchanging stability and instability.

6. The fixed point is given by $x_0 = \frac{\mu-1}{\mu}$ and the stability is given by $f'_\mu(x_0) = 2 - \mu$. So for $\mu = 3$, $|f'_\mu(x_0)| = 1$, $|f'_\mu(x_0)| < 1$ for $\mu < 3$, $|f'_\mu(x_0)| > 1$ for $\mu > 3$ giving the required change of stability. Note $f^2_\mu(x) = \mu^2 x(1 - x)(1 + \mu x(x - 1))$. So the fixed points of f^2_μ given by $f^2_\mu(x) = x$ are $x = 0$ and $x = \frac{\mu-1}{\mu}$ (which are the fixed points of f_μ) and two further fixed points $x = \mu + \mu^2 \pm \mu\sqrt{(\mu^2 - 2\mu - 3)}$ which exist for $\mu > 3$ to form a period-2 orbit.

7. Note $\cos(2\theta) = -1 + 2\cos^2(\theta)$ and consider the substitution $x = \cos(\theta)$ to give a conjugacy between f and the map $\theta \mapsto 2\theta$. The map g is conjugate $\theta \mapsto 3\theta$ noting the formula for $\cos(3\theta)$. Similarly, the polynomial for $\cos(m\theta)$ in terms of $\cos(\theta)$ provides a conjugate polynomial for a map on m-symbols.

8. A uniform density for a map f on $\mathbb{I}$ means that for increasingly large numbers of iterates n of f, the expected number which lies in the interval (a, b) will be in proportion to its relative length in $\mathbb{I}$, i.e. $b - a$.

9. $f(x) = x \implies g(x) = 0$ if $g'(x) \neq 0$. Let $x = x^*$ be a zero of the map g and $x_n \approx x^*$. Then $0 = g(x^*) = g(x_n) + (x_n - x^*)g'(x_0) + (x_n - x^*)^2 g''(\xi_n)/2$ for some $\xi \in [x^*, x_n)$ using a Taylor expansion

with Lagrange remainder. Rearranging, and using the notation of the N–R iteration rule, gives

$$(x_n - x^*) - \frac{g(x_n)}{g'(x_n)} = -\frac{(x_n - x^*)^2 g''(\xi_n)}{2g'(x_n)} \qquad (5.1)$$

$$\implies |x_{n+1} - x^*| = (x_n - x^*)^2 \frac{|g''(\xi_n)|}{2|g'(x_n)|}. \qquad (5.2)$$

10. $x_0 = 1$; $x_1 = 3$; $x_2 = 2.333$; $x_3 = 2.2381$.

References

1. M. Barenco and D. K. Arrowsmith, The autocorrelation of double intermittency maps and the simulation of computer packet traffic, *Dynam. Syst.* **19**(1), 61–74 (2004).
2. P. Embrechts, *Selfsimilar Processes*. Princeton University Press (2009).
3. R. L. Devaney, *An Introduction to Chaotic Dynamical Systems*. Benjamin-Cummings (1986).
4. G. H. Hardy, *An Introduction to the Theory of Numbers*. Oxford University Press (1979).
5. H. Nagashima and Y. Baba, *Introduction to Chaos*. IOP (1999).
6. Y. Pomeau and P. Manneville, Intermittent transition to turbulence in dissipative dynamical systems, *Commun. Math. Phys.* **74**, 189–197 (1980).
7. H. G. Schuster and W. Just, *Deterministic Chaos: An Introduction*. Wiley-VCH, Berlin (2005).
8. G. Setti, R. Rovatti and G. Mazzini , *Chaos-Based Generation of Artificial Self-Similar Traffic*. Springer, Berlin, pp. 159–190 (2005).

* 9 7 8 1 7 8 6 3 4 1 0 2 0 *